Exploring HAA Formation Pathways During Chloramination

Awwa Research Foundation

About the Awwa Research Foundation

The Awwa Research Foundation (AwwaRF) is a member-supported, international, nonprofit organization that sponsors research to enable water utilities, public health agencies, and other professionals to provide safe and affordable drinking water to consumers.

The Foundation's mission is to advance the science of water to improve the quality of life. To achieve this mission, the Foundation sponsors studies on all aspects of drinking water, including supply and resources, treatment, monitoring and analysis, distribution, management, and health effects. Funding for research is provided primarily by subscription payments from approximately 1,000 utilities, consulting firms, and manufacturers in North America and abroad. Additional funding comes from collaborative partnerships with other national and international organizations, allowing for resources to be leveraged, expertise to be shared, and broad-based knowledge to be developed and disseminated. Government funding serves as a third source of research dollars.

From its headquarters in Denver, Colorado, the Foundation's staff directs and supports the efforts of more than 800 volunteers who serve on the board of trustees and various committees. These volunteers represent many facets of the water industry, and contribute their expertise to select and monitor research studies that benefit the entire drinking water community.

The results of research are disseminated through a number of channels, including reports, the Web site, conferences, and periodicals.

For subscribers, the Foundation serves as a cooperative program in which water suppliers unite to pool their resources. By applying Foundation research findings, these water suppliers can save substantial costs and stay on the leading edge of drinking water science and technology. Since its inception, AwwaRF has supplied the water community with more than $300 million in applied research.

More information about the Foundation and how to become a subscriber is available on the Web at **www.awwarf.org**.

Exploring HAA Formation Pathways During Chloramination

Prepared by:
Tanju Karanfil, **Ying Hong**, **Hocheol Song**, and **Olivia Orr**
Clemson University
Department of Environmental Engineering and Science
L.G. Rich Environmental Research Laboratory
Box 340919
Clemson, SC 29634-0919

Sponsored by:
Awwa Research Foundation
6666 West Quincy Avenue, Denver, CO 80235-3098

Published by:

Awwa Research Foundation

American Water Works Association

IWA Publishing

DISCLAIMER

This study was funded by the Awwa Research Foundation (AwwaRF). AwwaRF assumes no responsibility for the content of the research study reported in this publication or for the opinions or statements of fact expressed in the report. The mention of trade names for commercial products does not represent or imply the approval or endorsement of AwwaRF. This report is presented solely for informational purposes.

Printed in the U.S.A.

Printed on recycled paper

CONTENTS

TABLES

FIGURES

FOREWORD

The Awwa Research Foundation (AwwaRF) is a nonprofit corporation that is dedicated to the implementation of a research effort to help utilities respond to regulatory requirements and traditional high-priority concerns of the industry. The research agenda is developed through a process of consultation with subscribers and drinking water professionals. Under the umbrella of a Strategic Research Plan, the Research Advisory Council prioritizes the suggested projects based upon current and future needs, applicability, and past work; the recommendations are forwarded to the Board of Trustees for final selection. The foundation also sponsors research projects through an unsolicited proposal process; the Collaborative Research, Research Applications, and Tailored Collaboration programs; and various joint research efforts with organizations such as the U.S. Environmental Protection Agency, the U.S. Bureau of Reclamation, and the Association of California Water Agencies.

This publication is a result of one of these sponsored studies, and it is hoped that its findings will be applied in communities throughout the world. The following report serves not only as a means of communicating the results of the water industry's centralized research program but also as a tool to enlist the further support of the nonmember utilities and individuals.

Projects are managed closely from their inception to the final report by the foundation's staff and large cadre of volunteers who willingly contribute their time and expertise. The foundation serves a planning and management function and awards contracts to other institutions such as water utilities, universities, and engineering firms. The funding for this research effort comes primarily from the Subscription Program, through which water utilities subscribe to the research program and make an annual payment proportionate to the volume of water they deliver and consultants and manufacturers subscribe based on their annual billings. The program offers a cost-effective and fair method for funding research in the public interest.

A broad spectrum of water supply issues is addressed by the foundation's research agenda: resources, treatment and operations, distribution and storage, water quality and analysis, toxicology, economics, and management. The ultimate purpose of the coordinated effort is to assist water suppliers to provide the highest possible quality of water economically and reliably. The true benefits are realized when the results are implemented at the utility level. The foundation's trustees are pleased to offer this publication as a contribution toward that end.

David E. Rager
Chair, Board of Trustees
Awwa Research Foundation

Robert C. Renner, P.E.
Executive Director
Awwa Research Foundation

ACKNOWLEDGMENTS

The authors would like to express their appreciation to various parties that assisted in different aspects of the project. The technical and administrative staff of Greenville Water System, Myrtle Beach and Spartanburg R.B. Simms water treatment plants assisted in collection and isolation of dissolved organic matter samples used during the project.

The Project Advisory Committee (PAC) consisted of Stuart Krasner of Metropolitan Water District of Southern California, Richard Miltner of the United States Environmental Protection Agency, and Peter Vikesland of Virginia Polytechnic Institute. PAC members provided valuable comments during the course of the project. Alice Fulmer of AwwaRF served as the project officer and facilitated successful completion of the project.

The Department of Environmental Engineering and Science at Clemson University provided the analytical facilities for the project.

EXECUTIVE SUMMARY

Disinfection by-products (DBPs) are formed as a result of reactions between naturally occurring dissolved organic matter (DOM) and oxidants/disinfectants such as chlorine. The presence of DBPs in drinking water is undesirable because they are suspected to be toxic, carcinogenic, and mutagenic to humans if ingested over extended periods of time. Therefore, the United States Environmental Protection Agency (USEPA) promulgated a two-stage Disinfectants/Disinfection By-Products Rule (D/DBPR) in December 1998 to minimize the risk of DBPs in drinking water to public health. Stage 2 of the D/DBPR, promulgated in January 2006, requires water utilities to comply with the maximum contaminant levels (MCLs) of 80 μg/L of total trihalomethanes (THM, which includes chloroform, bromodichloromethane, dibromochloromethane, and bromoform) and 60 μg/L of five haloacetic acids (HAA5, which is the sum of mono-, di-, and trichloroacetic acids, and mono- and dibromoacetic acids) at each individual monitoring location in a distribution system in the next six years. In order to comply with increasingly stringent regulations, drinking water utilities have been exploring various strategies to minimize DBP formation.

The use of chloramine for DBP control has received increasing attention in the water treatment industry since the first discovery of THM in drinking waters. The survey results of 543 utilities reported by McGuire and Meadow in 1988 showed that 100 water utilities (i.e., 18% of the participants) have switched from chlorination to chloramination since 1979. In another major survey conducted in 1998, it was found that 29% of water utilities have been using chloramine for secondary disinfection. The most recent survey of 363 drinking water utilities across all 50 states in 2005 showed that 68% and 29% use chlorine and chloramines as secondary disinfectant, respectively, and 3% does not use any disinfectant. In this latest survey, it was also found that of those 257 utilities using chlorine or no secondary disinfectant, seven intend to convert to chloramines, and another seventy seven are considering switching to chloramines with the onset of the implementation of Stage 2 D/DBPR.

Chloramination can almost completely suppress THM formation and reduce HAA concentration to 3–30% of the amounts formed during chlorination, being a viable alternative to water utilities for complying with the Stage 2 D/DBPR. This is the main reason for the increased popularity of chloramination in drinking water treatment industry in the US in recent years. However, chloramination has some disadvantages: (1) it is a relative weaker disinfectant and oxidant than free chlorine, requiring much higher concentration (C) and longer contact time (T) for pathogen inactivation and oxidation as compared to chlorine (which is why it is typically used as a secondary disinfectant and not as the primary disinfectant/oxidant), (2) it may cause nitrification in distribution systems, (3) some DBPs (nitrosamines, cyanogen chloride) that are not currently regulated may also form, and (4) it may trigger lead dissolution in some distribution systems.

The main objectives of this study were to investigate the pathways of dihalogenated acetic acids (DXAA), the major class of HAA forming during chloramination, and to systematically examine the roles of important factors, namely DOM characteristics, pH and bromide, in DXAA formation and speciation. Understanding the pathways of DXAA formation and speciation during chloramination is critical for developing strategies toward their control and modeling of their formation through mechanistic or empirical models.

RESEARCH APPROACH

DOM from the influents of three surface water treatment plants in South Carolina with a wide range of organic matter characteristics, Myrtle Beach (MB) (Specific Ultraviolet Absorbance at 254 nm wavelength ($SUVA_{254}$) = 4.9 L/mg-m), Spartanburg (SP) ($SUVA_{254}$ = 3.2 L/mg-m), and Greenville (GV) ($SUVA_{254}$ = 1.9 L/mg-m) were collected using a field-scale reverse osmosis (RO) system. The RO concentration allowed the project team to perform experiments at the same dissolved organic carbon (DOC) concentrations, pH, and bromide levels for waters collected from different sources. The RO concentration also allowed fractionation of the DOM using the same initial concentration (C_0) and column capacity factor (k') in the subsequent resin fractionation experiments. Kinetics experiments were performed using the three RO concentrates and three DOM fractions obtained from two source waters (GV and MB) at three pH (6, 7.5 and 9) and three bromide levels (i.e., Br/DOC ~50, 150 and 300 μg/mg) to examine the formation and speciation of HAA during chloramination. Furthermore, it was found during the course of the study that the quenching method can have important consequences on the determination of HAA formation kinetics during chloramination. A limited number of experiments were performed to examine the role of quenching method in HAA determination.

CONCLUSIONS AND RECOMMENDATIONS

The important conclusions and recommendations from this project are summarized below:

HAA Formation Pathways and Kinetics During Chloramination

- It was found that the direct reaction of NH_2Cl with DOM is the major pathway for HAA formation during chloramination. Experiments performed in the presence (where NH_2Cl decomposition and $NHCl_2$ formation are significantly suppressed) and absence (where NH_2Cl decomposition can occur) of background ammonia indicated that a major amount (~80%) of DXAA was formed from the direct reaction(s) of NH_2Cl with DOM at pH 6. The remaining ~20% resulted from reactions with HOCl. Results showed that $NHCl_2$ did not form HAA under typical drinking water conditions.
- It was shown that HAA formation kinetics during chloramination are not extremely fast, as has been previously reported in the literature. It was discovered and reported for the first time that the main reason for the different patterns reported for HAA formation kinetics in the literature is the use of NH_4Cl as the quenching agent recommended by USEPA Method 552.3. Since NH_4Cl does not quench residual chloramine, the reactions of chloramine can continue until the samples are analyzed. As a result, false high HAA concentrations are measured at early reaction periods. For one water tested in this study, it was found that for the samples within the first 8 hr of chloramination at pH 7 and 8.3, a positive error of up to 10 μg/L was observed in the measured HAA concentrations by using NH_4Cl as the quenching agent and storing the samples in a refrigerator for seven days. The magnitude of the error decreased with increasing contact time.
- In a limited number of experiments, five quenching approaches, stoichiometric amount of Na_2SO_3, NH_4Cl, sodium meta-arsenite ($NaAsO_2$), increasing pH to 10 using a

borate acid buffer solution and increasing pH to 10 followed by 100 mg/L ascorbic acid (i.e., Vitamin C) addition, were tested side by side using the same chloraminated sample. $NaAsO_2$ produced the best results, showing no effect on all nine HAA determinations even after 14 days of storage. Furthermore, the biocide capability of $NaAsO_2$ has been previously shown for distribution system samples. Unfortunately, due its toxic nature, the use of $NaAsO_2$ in practical application poses various hazard and waste handling issues. More research is needed to identify and validate a proper quenching approach for chloraminated DBP samples in practical applications. If the residual chloramine concentration can be quickly determined in a sample, the use of a stoichiometric amount of Na_2SO_3 can be considered. However, it should be noted that Na_2SO_3 does not work as a biocide, thus degradation of HAA species may occur in biologically active samples. The use of excess Na_2SO_3 concentrations results in the decomposition of some brominated HAA species.

- The results also showed that chloramination is an effective way to control DBP formation, especially by minimizing the lag time between chlorine and ammonia addition and maintaining pH above 7.5. Since HAA formation in chloraminated samples was found to occur mainly from the direct reaction of NH_2Cl with DOM, the enhanced removal of DOM can decrease HAA formation during chloramination.

HAA Formation From DOM and DOM Fractions During Chloramination

- The results obtained in this study and in previous investigations with preformed chloramines indicate that at least 80% of HAA to be formed from chloramines(s) in natural waters will be in the DXAA form, and the contribution of DXAA may reach and exceed over 90% in some cases. Small amount of monohalogenated haloacetic acids (MXAA) formation may also be observed, while formation of trihalogenated haloacetic acids (TXAA) is usually negligible.
- *pH effect:* DXAA formation both on a molar and mass concentration basis decreased with increasing pH at all three bromide concentrations in all three waters. Formation of MXAA, although in very small amounts, appeared to decrease with increasing pH, while TXAA formation was usually negligible, except small amounts of TXAA formation was observed at pH 6 with the highest Br/DOC ratios. These trends were different from those observed during chlorination. For chlorination, formation of MXAA and DXAA is almost independent of pH at typical water treatment conditions (i.e., pH 6 to 9), while formation of TXAA decreases with increasing pH. The formation of DXAA was sensitive to pH during chloramination. However, the pH effect was not proportional to the change in pH. There was a larger decrease in DXAA yield when pH was increased from 6 to 7.5 than from 7.5 to 9.0.
- *Bromide effect:* Bromide effect was pH dependent. At pH 6, DXAA formation increased and shifted toward brominated species with increasing bromide concentration. Small amounts of MXAA formation were observed, which shifted also from MCAA to MBAA with increasing Br/DOC ratio. The presence of bromide resulted in TXAA formation at the highest Br/DOC ratios for both DOMs. At pH 7.5, the impact of bromide on DXAA formation and NH_2Cl decomposition was relatively small, and at pH 9, it was essentially negligible. Bromine incorporation was higher and occurred more rapidly at pH 6 than pH 7.5. No brominated species were formed at pH 9.

- *DOM effect:* The aromatic content of DOM (i.e., $SUVA_{254}$) appeared to have some impact on the DXAA yields during chloramination at pH values (e.g., pH 6 tested in this study) that considerably increase NH_2Cl decomposition. However, the effect of $SUVA_{254}$ on the HAA yield was much less pronounced during chloramination than chlorination. At pH 6 the formation of DXAA in MB DOM ($SUVA_{254}$ = 4.9 mg/L-m) was about 10–30% higher (on both a mass and molar basis) than in GV ($SUVA_{254}$ = 2.6 mg/L-m) and SP DOM ($SUVA_{254}$ = 3.2 mg/L-m) at three bromide levels. At pH 7.5 and 9.0, both GV and MB DOMs showed very similar DXAA yields. These results indicated that pH was a more important factor controlling the extent of DXAA formation and speciation than $SUVA_{254}$ of DOM and bromide in typical drinking water conditions.
- *Reactivity of DOM fractions:* To investigate the effect of DOM characteristics, two surface waters with different $SUVA_{254}$ values were fractionated into hydrophobic (HPO), transphilic (TPH), and hydrophilic (HPI) fractions. Generally, with the decrease in pH, the effect of DOM characteristics and bromide became more important. At pH 9, the effect of DOM was minimal and the effect of bromide was not observed. At pH 7.5, both effects were observed but not to a significant degree. At pH 6, DXAA yield increased with increasing $SUVA_{254}$ for all fractions. However, the effect of $SUVA_{254}$ on the HAA yields was less pronounced during chloramination than during chlorination. The difference in DXAA yields between chlorination and chloramination increased with increase in $SUVA_{254}$. The overall trend of DXAA formation was (MBHPO > GVHPO) > (MBTPH ≈ MBHPI) > (GVTPH ≈ GVHPI), indicating that the hydrophobic components were the major precursors of DXAA. The opposite trend held true for the bromine incorporation factor (n): GVHPI ≈ GVTPH > MBHPI > MBTPH ≈ GVHPO > MBHPO. The bromine incorporation factor showed an inverse but not a linear correlation with $SUVA_{254}$.

RELEVANCE TO DRINKING WATER UTILITIES

Since it was found that the majority of HAA are formed as a result of the direct reaction of NH_2Cl with DOM, even if NH_2Cl decomposition is minimized or if NH_2Cl is formed by mixing Cl_2 and NH_3 at the same injection location, an appreciable amount of DXAA may still form during chloramination. It was shown that HAA formation decreases with increasing pH. However, the pH effect was not proportional to the change in pH. There was a larger decrease in DXAA yield when pH was increased from 6 to 7.5 than from 7.5 to 9.0. The water utilities intending to decrease HAA formation using chloramination should maintain pH above 7.5 and minimize the lag time between chlorine and ammonia addition or add ammonia before chlorine. One additional advantage of having an operational pH of 7.5 or above for chloramination is to reduce the formation of brominated HAA species.

Although the HPI fractions were more reactive with bromide than the HPO fractions, the HPO fractions consistently showed higher HAA yields than the TPH and HPI fractions. Thus, the enhanced removal (e.g., (enhanced) coagulation/flocculation) of the hydrophobic DOM component especially in moderate and high $SUVA_{254}$ waters can assist with the better control of DBP formation during chloramination.

It was shown that HAA formation kinetics during chloramination are not extremely fast, as has been previously reported in the literature. Therefore, HAA formation can continue to occur after the water leaves a treatment plant. The impact of HAA formation kinetics depends also on chloramination practice (i.e., chlorine/ammonia addition sequence).

Since NH_4Cl recommended by USEPA Method 552.3 does not quench residual chloramine, there is a possibility of overestimating HAA formation for distribution system samples close to the treatment plant. The magnitude of the error decreases with increasing distribution system residence time. Among five quenching approaches, stoichiometric amount of Na_2SO_3, NH_4Cl, sodium meta-arsenite ($NaAsO_2$), increasing pH to 10 using a borate acid buffer solution and increasing pH to 10 followed by 100 mg/L ascorbic acid (i.e., Vitamin C) addition, tested, $NaAsO_2$ produced the best results, showing no effect on all nine HAA determinations after 14 days of storage. Furthermore, the biocide capability of $NaAsO_2$ has been previously shown for distribution system samples. Unfortunately, due its toxic nature, the use of $NaAsO_2$ in practical application poses various hazard and waste handling issues.

FUTURE RESEARCH NEEDS

The experiments in this study were performed using the water samples collected from the influents of three water treatment plants. Future studies should be conducted using the water samples after treatment processes prior to any disinfectant addition and water distribution system samples to further confirm and expand the findings regarding the HAA formation routes and quenching agent effect, respectively. More research is needed to identify and validate a proper quenching approach for chloraminated DBP samples in practical applications. There is wealth of HAA formation kinetics data obtained in this study that can be modeled with the DBP formation models that were developed for chloramination in the recent years. Furthermore, new and improved mechanistic HAA formation models can be developed using the information about the HAA formation routes obtained in this study.

CHAPTER 1
INTRODUCTION, BACKGROUND, AND PROJECT OBJECTIVES

INTRODUCTION

The use of chloramination in drinking water treatment has gained increasing popularity in the United States during the last two decades, primarily due to the increasingly stringent regulations imposed by the United States Environmental Protection Agency (USEPA) under the Disinfectants/Disinfection Byproducts Rule (D/DBPR). Stage 1 of the D/DBPR set the maximum contaminant levels (MCLs) of total trihalomethanes (THMs: chloroform, bromodichloromethane, dibromochloromethane, and bromoform) and five haloacetic acids (HAA5: the sum of mono-, di-, and trichloroacetic acids, and mono- and dibromoacetic acids) at 80 and 60 μg/L, respectively, for the running annual average of a water distribution system (USEPA 1998). Stage 2 of the D/DBPR, promulgated in January 2006, requires water utilities to comply with the same MCLs at each individual monitoring location in a distribution system (i.e., locational running annual averages) (USEPA 2006).

Several surveys characterizing the use of disinfectants in drinking water utilities have been conducted in the United States (US). A survey of 543 utilities conducted by McGuire and Meadow (1988) showed that 100 water utilities (i.e., 18% of the participants) switched from chlorination to chloramination since the promulgation of the THM Rule (THM MCL = 100 μg/L) in 1979. In another major survey conducted in 1998, it was found that 29% of medium and large water utilities used chloramines (American Water Works Association Committee Report 2000). The most recent survey of 363 drinking water utilities across all 50 states showed that 68% and 29% used chlorine and chloramines as a secondary disinfectant, respectively, and 3% did not use any secondary disinfectant (Seidel et al., 2005). In this survey, it was also found that of those 257 utilities using chlorine or no secondary disinfectant, seven planned to convert to chloramines, and another seventy seven were considering switching to chloramines due to the implementation of the Stage 2 D/DBPR.

While it appears that chloramine use will further increase among water utilities in the coming years, it has some disadvantages: (1) it is a relative weaker disinfectant and oxidant than free chlorine, requiring much higher concentration (C) and longer contact time (T) for pathogen inactivation and oxidation as compared to chlorine (which is why it is typically used as a secondary disinfectant and not as the primary disinfectant/oxidant), (2) it may cause nitrification in distribution systems, (3) some DBPs (nitrosamines [Mitch et al. 2003], cyanogen chloride [Krasner et al. 1989]) that are not currently regulated may also form, and (4) it may trigger lead dissolution in some distribution systems (Edwards and Dudi 2004).

Despite these disadvantages, chloramination is used by water utilities because it almost completely suppresses THM formation and reduces HAA concentrations to 3–30% of the amounts formed during chlorination (Long et al. 1992, Cowman and Singer 1996, Symons et al. 1998, Zhang et al. 2000). Dihalogenated haloacetic acids (DXAAs) are the primary regulated DBP species formed during chloramination, whereas the formation of monohalogenated (MXAAs) and trihalogenated haloacetic acids (TXAAs) is significantly suppressed (Symons et al. 1998, Singer et al. 1999, Diehl et al. 2000). Although chloramination does not completely eliminate DXAA formation, it can provide a viable alternative for water utilities to comply with the

Stage 2 D/DBPR. This is the main reason for the increased popularity and use of chloramination by the drinking water treatment industry in the US in recent years.

The main objectives of this study were to investigate the pathways of DXAA, the major class of HAA forming during chloramination, and to systematically examine the roles of important factors, namely dissolved organic matter (DOM) characteristics, pH and bromide, in DXAA formation and speciation. Understanding the pathways of DXAA formation and speciation during chloramination is critical for developing strategies toward their control and modeling of their formation through mechanistic or empirical models.

BACKGROUND

Chloramine Chemistry in the Absence of Bromide

Chloramination is the process that uses mainly monochloramine (NH_2Cl) for primary or secondary disinfection. When free chlorine (HOCl) and ammonia (NH_3) are added to water, a series of reactions can occur, forming different chloramine species: NH_2Cl, dichloramine ($NHCl_2$), and trichloramine (NCl_3). NH_2Cl and $NHCl_2$ constitute the combined residual chlorine in water. NH_2Cl is the most desired and commonly used form of chloramines in drinking water treatment. However, NH_2Cl is inherently unstable, decomposing as pH decreases, even in the absence of reactive background organic or inorganic compounds, a process called autodecomposition. Margerum and Valentine and their coworkers have extensively studied the complex chloramine chemistry in order to identify the most important reactions. Their results identified three major groups of reactions (Table 1.1) for NH_2Cl autodecomposition (Duirk 2003): (i) substitution/hydrolysis reactions (reactions 1-1 to 1-4) (ii) disproportionation reactions (reactions 1-5 and 1-6), and (iii) redox reactions (reactions 1-7 to 1-10) occurring in the absence of free chlorine. Equilibrium reactions for chlorine, ammonia and carbonate species (reactions 1-11 to 1-14) are also included in Table 1.1.

Substitution/hydrolysis (reactions 1-1 to 1-4) and disproportion (reactions 1-5 and 1-6) are the major reactions that occur during chloramination. When chlorine is dispersed in water, a rapid and complete hydrolysis occurs, forming HOCl, which is a weak acid with a pK_a of 7.5. Therefore, the distribution of HOCl and the hypochlorite ion (OCl^-) in water is dependent on pH. The nonionic forms of HOCl and NH_3 favor NH_2Cl formation. The optimum pH to form NH_2Cl (reaction 1-1) is the midpoint between the two pK_a values of reactions 1-11 and 1-12, approximately pH 8.5 (Haas 1999). The rate of NH_2Cl formation is so fast in the range of pH 7 to 8.5 that the time to convert 99% of HOCl to NH_2Cl is less than one second (White 1999). The rate of the reaction slows down at lower and higher pH values. $NHCl_2$ and NCl_3 formation in drinking water is undesirable due to odor and taste problems (Krasner and Barrett 1985). NCl_3 formation reaction, which is negligible under typical drinking water conditions, is not listed in Table 1.1.

These competing reactions are primarily dependent on pH and the chlorine to nitrogen (Cl_2/N) ratio. NH_2Cl is the dominant form of chloramines when the Cl_2/N weight ratio is less than 5:1, and the pH is approximately 8, representing the common conditions for drinking water chloramination. Current practice is to use Cl_2/N ratios from 3:1 to 5:1, typically 4:1 (USEPA 1999). A ratio ~4.5:1 is often used to minimize the formation of $NHCl_2$ and DBPs, as well as to control nitrification and biofilm growth associated with excess NH_3. $NHCl_2$, which becomes significant as the Cl_2/N ratio increases and pH decreases, can be formed either through the substitution reaction in the presence of HOCl (reaction 1-3) and/or the disproportionation reaction (reaction 1-5). Valentine

Table 1.1
Formation and decomposition reactions of chloramines in the absence of bromide

	Reaction stoichiometry	Rate/equilibrium constant (25°C)	Reference
1-1	$HOCl + NH_3 \rightarrow NH_2Cl + H_2O$	$k_1 = 1.5 \times 10^{10}\ M^{-1}h^{-1}$	Morris and Isaac 1981
1-2	$NH_2Cl + H_2O \rightarrow HOCl + NH_3$	$k_2 = 7.6 \times 10^{-2}\ h^{-1}$	Morris and Isaac 1981
1-3	$HOCl + NH_2Cl \rightarrow NHCl_2 + H_2O$	$k_3 = 1.0 \times 10^{6}\ M^{-1}h^{-1}$	Margerum et al. 1978
1-4	$NHCl_2 + H_2O \rightarrow HOCl + NH_2Cl$	$k_4 = 2.3 \times 10^{-3}\ h^{-1}$	Margerum et al. 1978
1-5	$NH_2Cl + NH_2Cl \rightarrow NHCl_2 + NH_3$	k_5: pH dependent*	Vikesland et al. 2001
1-6	$NHCl_2 + NH_3 \rightarrow NH_2Cl + NH_2Cl$	$k_6 = 2.16 \times 10^{8}\ M^{-2}h^{-1}$	Hand and Margerum 1983
1-7	$NH_2Cl + NHCl_2 \rightarrow N_2 + 3H^+ + 3Cl^-$	$k_7 = 55.0\ M^{-1}h^{-1}$	Leao 1981
1-8	$NHCl_2 + H_2O \rightarrow I^{\dagger} + 2HCl$	$k_8 = 4.0 \times 10^{5}\ M^{-1}h^{-1}$	Jafvert and Valentine 1987
1-9	$I^{\dagger} + NHCl_2 \rightarrow HOCl + N_2 + HCl$	$k_9 = 1.0 \times 10^{8}\ M^{-1}h^{-1}$	Leao 1981
1-10	$I^{\dagger} + NH_2Cl \rightarrow N_2 + H_2O + HCl$	$k_{10} = 3.0 \times 10^{7}\ M^{-1}h^{-1}$	Leao 1981
1-11	$HOCl \leftrightarrow H^+ + OCl^-$	$pK_a = 7.5$	Snoeyink and Jenkins 1980
1-12	$NH_4^+ \leftrightarrow H^+ + NH_3$	$pK_a = 9.3$	Snoeyink and Jenkins 1980
1-13	$H_2CO_3 \leftrightarrow H^+ + HCO_3^-$	$pK_a = 6.3$	Snoeyink and Jenkins 1980
1-14	$HCO_3^- \leftrightarrow H^+ + CO_3^{2-}$	$pK_a = 10.3$	Snoeyink and Jenkins 1980

Source: Duirk 2003.
* $k_5 = k_{H^+}[H^+] + k_{H_2CO_3}[H_2CO_3] + k_{HCO_3^-}[HCO_3^-]$: $k_{H^+} = 2.5 \times 10^7 M^{-2}h^{-1}$, $k_{H_2CO_3} = 4.0 \times 10^3 M^{-2}h^{-1}$, and $k_{HCO_3^-} = 800\ M^{-2}h^{-1}$ at 25°C.
† I: reactive intermediate.

and Jafvert (1988) demonstrated that the disproportionation reaction is more sensitive to pH than the substitution reaction, and the former reaction is the dominant pathway forming $NHCl_2$ at pH levels far below the pK_a of NH_3 ($pK_a = 9.3$) and in the presence of high background NH_3.

Because $NHCl_2$ is unstable and decomposes rapidly via base-catalyzed reactions at neutral or alkaline pH levels, it is responsible for total chlorine residual loss (reactions 1-7 to 1-9). Researchers have shown some evidence of the presence of a certain intermediate (I), which is released from $NHCl_2$ hydrolysis and decays almost instantaneously (Leao 1981, Hand and Margerum 1983, Valentine et al. 1986, Leung and Valentine 1994). The end products included nitrogen gas (N_2) and NH_3. The mechanism of $NHCl_2$ decay is still not completely understood.

Chloramine Chemistry in the Presence of Bromide

Chloramine chemistry is complicated in the presence of bromide (Br^-), which is widespread in source waters due to river water mixing with sea water, salt water intrusion into groundwater, or the dissolution of bromide containing minerals (Siddiqui and Amy 1993, Krasner et al. 1994). Br^- concentrations in natural waters in the US ranged from nondectable to 2.2 mg Br^-/L, with an average concentration of 100 μg/L, according to a survey conducted by Amy et al. (1994), and 70 μg/L based on an American Water Works Association Committee Report (2000). The

Table 1.2
Formation and decomposition reactions of chloramines in the presence of bromide

Reaction stoichiometry		Rate/equilibrium constant (25°C)	Reference
2-1	$NH_2Cl + H^+ \leftrightarrow NH_3Cl^+$	$K_e = 28\ M^{-1}$	Gray et al. 1978
2-2	$NH_3Cl^+ + Br^- \rightarrow NH_3Br^+ + Cl^-$	$k_{Br1} = 1.8 \times 10^8\ M^{-1}h^{-1}$	Trofe et al. 1980
2-3	$NH_3Br^+ \leftrightarrow NH_2Br + H^+$	$pK_a = 6.4$	Bousher et al. 1989
2-4	$NH_2Cl + NH_3Br^+ \rightarrow NHBrCl + NH_4^+$	k_{Br2}: fast	Vikesland et al. 2001
2-5	$HOCl + Br^- \rightarrow HOBr + Cl^-$	$k_{Br3} = 5.6 \times 10^6\ M^{-1}h^{-1}$	Bousher et al. 1986
2-6	$HOBr \leftrightarrow OBr^- + H^+$	$pK_a = 8.8$	Gazda and Margerum 1994
2-7	$HOBr + NH_2Cl \rightarrow NHBrCl + H_2O$	$k_{Br4} = 1.0 \times 10^9\ M^{-1}h^{-1}$	Gazda and Margerum 1994
2-8	$NHBrCl \leftrightarrow NBrCl^- + H^+$	$pK_a = 4.9$	Bousher et al. 1989
2-9	$HOBr + NH_3 \rightarrow NH_2Br + H_2O$	$k_{Br5} = 2.7 \times 10^{11}\ M^{-1}h^{-1}$	Wajon and Morris 1980
2-10	$NHBrCl + NH_2Cl \rightarrow N_2 + 3H^+ + 2Cl^- + Br^-$	k_{Br7}: fast	Vikesland et al. 2001

Source: Duirk 2003.

typical Br^-/DOC (dissolved organic carbon) ratio in surface waters is below 100 μg/mg, however, a ratio as high as 528 μg/mg was also reported (Summers et al. 1996).

The presence of bromide catalyzes the decomposition of NH_2Cl, resulting in a higher NH_2Cl demand. Important reactions associated with bromide are listed in Table 1.2. Reactions 2-1 to 2-4 describe the pathway of bromochloramine (NHBrCl) formation through bromide oxidation by NH_2Cl via the formation of the monochlorammonium ion (NH_3Cl^+), which is an acid catalyzed reaction (reaction 2-1). Once NH_3Cl^+ forms, the bromide ion is oxidized to NH_3Br^+ and then forms NHBrCl rapidly. Reactions 2-5 to 2-7 illustrate NH_2Cl reacting with hypobromous acid (HOBr) to form NHBrCl. At alkaline pH (i.e., above pH 8), reaction 2-1 (formation of NH_3Cl^+) and reaction 2-5 (formation of HOBr) are suppressed significantly, and so are the subsequent reactions. As a result, the bromide effect on NH_2Cl decay and DBP formation decreases significantly at pH levels above 8. Because the formation of dibromamine ($NHBr_2$) and tribromamine (NBr_3) are not insignificant at typical drinking water conditions (Duirk 2003), they are not included in Table 1.2.

HOBr is a very strong halogen substitution agent. HOBr is 10–25 times stronger than HOCl for halogen substitution during THM and HAA formation (Amy et al. 1991, Cowman and Singer 1996, Chang et al. 2001). Gazda et al. (1993) found that the reaction rate of HOBr with NH_2Cl (reaction 2-7) is at least 1000 times faster than that of Br^- with NH_2Cl at pH 6.5 (reactions 2-1 to 2-4). On the other hand, Valentine et al. (1998) suggested that HOBr occurrence was suppressed because the limited HOCl produced from NH_2Cl hydrolysis has a higher reaction rate with NH_3 rather than with Br^- or NH_2Cl. However, the formation of HOBr may not be completely negligible because the consequent reaction of HOBr and NH_2Cl is so fast that it can accelerate HOBr formation. Vikesland et al. (2001) reported that reactions 2-1 to 2-4 are usually the dominant reactions under drinking water conditions, but no data supporting this conclusion were provided.

It is evident that the presence of bromide significantly complicates the solution chemistry during chloramination. It was expected that more bromamines, chloramines and bromochloramines would co-exist in the chloramination system with bromide. However, Trofe et al. (1980) reported that bromochloramine was the principal product observed as NH_2Cl oxidizes bromide. The reaction was first order with respect to NH_2Cl, bromide, and hydrogen ion concentrations with a Cl_2/N molar ratio of 0.5:1 and a Br^-/NH_2Cl molar ratio more than 4:1 at pH 6.8 to 8.2. The results of later studies (Valentine et al. 1986, Bousher et al. 1989, Gazda et al. 1993) supported this finding. Valentine et al. (1986) reported that NHBrCl was the only oxidant besides NH_2Cl, with no bromamines being observed by using ultraviolet (UV) spectroscopy, an observation perhaps due to the specific experimental conditions: pH 6.5, low Cl_2/N ratio (1/6), Br^-/NH_2Cl molar ratio 20:1, and short reaction time (40 min). Gazda et al. (1993) used membrane introduction mass spectrometry in addition to UV spectroscopy to confirm the existence of NHBrCl. They reported that NHBrCl is the primary product of NH_2Cl reactions with bromide at Br^-/NH_2Cl molar ratios more than 12:1 in the pH range of 6.5 to 7. The formation of dibromochloramine (NBr_2Cl) was also observed. The formation of NBr_2Cl is not likely to occur in drinking water treatment conditions since the molar ratio of Br^-/NH_2Cl is less than one. Valentine (1986) further reported that NHBrCl reacted with N,N-diethyl-p-phenylenediamine (DPD) rapidly at pH 6.5, suggesting that the bromine atom of the NHBrCl is very labile and reactive. As the reaction 2-10 in Table 1.2 shows, NHBrCl undergoes its own auto-decomposition reaction leading to the ultimate regeneration of bromide, indicating that bromide primarily acts as a catalyst to increase the NH_2Cl decay rate (Vikesland et al. 2001).

Halogenated DBP Formation During Chloramination

In comparison to chlorination, chloramination decreases halogenated DBP formation significantly, including by-products such as total organic halogen (TOX), THMs, and HAAs (Symons et al. 1999). This is due to the weaker halogen substitution potential of chloramines than chlorine. However, some amount of DXAAs can still be formed (Cowman and Singer 1996) and cyanogen chloride formation is even higher as compared to chlorination (Krasner et al. 1989). Halogenated DBP formation during chloramination depends upon various factors, including pH, temperature, DOM characteristics, bromide concentration, chloramine application techniques, Cl_2/N ratio, and the reaction time. It should be noted that the review in the following section is for preformed chloramines or chloramines formed in-situ by addition of chlorine and ammonia at the same location. The issues for exceptional conditions (e.g., a lag time between chlorine and ammonia additions), when used in the review, will be clearly mentioned.

TOX Formation

In general, chloramines form halogenated organic by-products to a lesser degree than HOCl, with TOX representing the overall halogenated DBP formation. Being a weaker oxidant, NH_2Cl is less likely than HOCl to break down large DOM molecules into smaller fragments. It is hypothesized that most of the chlorinated organic by-products remain as high molecular weight compounds during chloramination (Jensen et al. 1985). However, the characteristics of these by-products are not well known. Currently identified DBPs constitute only a small fraction of TOX formed during chloramination. THM and HAA constitute less than 20% of TOX in chloramination (Symons et al. 1999, Wu et al. 2003) and almost 50% in chlorination (Christman et al. 1983, Singer and Chang 1989, Reckhow et al. 1990). The remaining DBPs still need to be identified and they may be, in fact, responsible for the most health concern (Johnson and Jensen 1986).

Although it is indicated that mutagenic activity observed in chloraminated water is typically less than that in chlorinated water (Noot et al. 1989), no definitive conclusions can be drawn from the literature and uncertainties about chloramine by-products still exist. Diehl et al. (2000) investigated the dissolved organic halide (DOX) formation of three natural waters at different pH, Cl_2/N ratios, and bromide concentrations and reported that the DOX formation decreased with increasing pH, decreasing Cl_2/N ratio, and decreasing bromide concentration.

THM Formation

Significant reduction of THM formation using chloramination has been recognized since early 1980s (Brodtmann and Russo 1979, Norman et al. 1980). However, the reason(s) behind this reduction are not known. Although very small amounts form, THM formation decreases with increasing pH during chloramination (Diehl et al. 2000, Wu et al. 2003). In the presence of bromide, THM formation increases and more brominated species are formed (Diehl et al. 2000, Wu et al. 2003).

HAA Formation

Only a few studies have examined the formation and speciation of HAAs, the main class of regulated DBPs formed during chloramination. HAAs are composed of MXAAs, DXAAs, and TXAAs. In chlorination, TXAAs and DXAAs dominate, and the relative amount of these two species depends on Cl_2/DOC ratio and DOM characteristics (Reckhow and Singer 1985, Reckhow et al. 1990). In chloramination (using preformed NH_2Cl), however, DXAA is by far the main class of HAA species. MXAAs and TXAAs usually account for a small fraction of total HAA formation. For example, the chloramination experiments conducted by Cowman and Singer (1996), using an XAD-8 resin isolate (i.e., hydrophobic fraction of DOM) from the Floridian aquifer at pH 8, showed that on a mole fraction basis, approximately 60% DXAAs, 33% MXAAs and less than 10% TXAAs were formed. On the other hand, Diehl et al. (2000) and Hwang et al. (2000) reported that almost all of the HAAs formed were DXAAs, with both TXAA and MXAA species being negligible. Recently, Speitel et al. (2004) reported that 72-hr DXAA formation represented at least 70% of HAA formation at pH 7 and 9 for different natural waters, while MXAA formation was also observed, but TXAA formation was, in general, negligible. Although minimal formation of TXAAs has been observed in many studies, Teng and Veenstra (1996) reported comparable levels of dichloroacetic acid (DCAA) and trichloroacetic acid (TCAA) formation during the chloramination of hydrophobic and hydrophilic fractions of a river water DOM.

Cowman and Singer (1996) noted that HAA formation patterns observed during chloramination mimic low-dose chlorination conditions, because lower Cl_2/DOC ratios preferentially favor DXAA formation. They postulated that the small amount of HOCl produced as a result of NH_2Cl hydrolysis (reaction 1-2 in Table 1.1) plays the major role in the preferential formation of DXAA over TXAA during chloramination. However, hydrolysis of NH_2Cl is a very slow reaction with a rate constant of 7.6×10^{-2} h^{-1}, with an extent of only 0.1–1% of the NH_2Cl at a neutral pH for NH_2Cl concentrations in the range of 1 to 10 mg/L (Morris and Isaac, 1981). It is possible that the rapid reaction of HOCl with DOM may drive the hydrolysis reaction forward to produce more HOCl. Zhang et al. (2000) made calculations to further examine this hypothesis, and found that halogenated aromatic organic matter (ClArOH) resulting from NH_2Cl hydrolysis could contribute to approximately 26% of total organic chlorine (TOCl) after a 5-day reaction time. Therefore, the small amount of HOCl in equilibrium with NH_2Cl may constitute an appreciable, but not the major pathway, for the formation of TOCl and DBPs during chloramination. Additional reactions that occur during chloramination can also contribute to HAA formation.

Although the reaction rate of NH_2Cl decay and the subsequent HAA yield are low as compared to HOCl, NH_2Cl is known to transfer active chlorine to the nitrogen of amines and amino acids, forming organic chloramines. It can also add chlorine to activated aliphatic carbon-carbon double bonds (Johnson and Jensen 1986). Under certain conditions, chlorine substitution into activated aromatic rings by NH_2Cl has been observed (Burttschell et al. 1959, Topudurti and Haas 1991, Heasley et al. 2004, Greyshock and Vikesland, 2006). Considering the much higher concentration of NH_2Cl than both HOCl and $NHCl_2$, it is expected that the direct reaction between NH_2Cl and DOM may result in considerable amount of HAA formation. Today, there is no information about the significance of this reaction in the literature. Only recently during the course of this project, Duirk (2003) reported the modeling of DCAA formation during chloramination by postulating both a fast direct reaction between NH_2Cl and DOM and a slow reaction between HOCl (formed as a result of NH_2Cl hydrolysis) and DOM, but the relative contribution of each reaction to overall HAA formation has not been evaluated.

The formation of $NHCl_2$ can be observed at low pH during chloramination. However, there is no information in the literature about the role of $NHCl_2$ in the HAA formation. Diehl et al. (2000) suggested that $NHCl_2$ or the decomposition product HOCl is the active halogenating agent at pH 6, while NH_2Cl may play a dominant role at pH 8, because $NHCl_2$ decomposition is base catalyzed. However, no experimental evidence was presented to support this hypothesis.

In this research, it is considered that there are three possible reactants that can interact with DOM forming HAA in the absence of bromide during chloramination: NH_2Cl, HOCl, and $NHCl_2$. The conceptual illustration of HAA formation pathways during chloramination is illustrated in Figure 1.1.

HAA Formation Kinetics

The chloramination research conducted within the past decade has focused primarily on HAA and THM formation over a 24-hour or longer reaction time. The formation and speciation kinetics of HAA have not been systematically investigated. Only a limited amount of information is available on HAA production kinetics during chloramination, with significantly different formation patterns, as shown in Figure 1.2.

Speitel and coworkers reported significantly fast HAA kinetics for Metedenock river water (at pH 8 under the uniform formation conditions (UFC) protocol), indicating that approximately 73% of the 72-hour DXAA (both chlorinated and brominated HAAs) formed after only 0.5 hours and almost 82% after 1 hour (Figure 1.2) (Pope et al. 2002, Pope et al., 2003, Speitel et al. 2004). They also observed similar rapid DXAA formation kinetics for softened Lake Austin water. Hua and Reckhow (2005a) also reported similar fast kinetics (85% of 72-hr yield formed in 30 min) during chloramination. These observations of extremely rapid formation rate are unexpected, because chlorine is a much stronger halogenation agent than chloramines, and about 80 to 90% of HAA formation was not completed during chlorination until after 24 hours (Krasner et al. 1996b, Singer et al. 1999). In addition, the reported rapid DCAA formation pattern did not follow the slow NH_2Cl decomposition patterns at pH 8. Singer et al. (1999) also reported a relatively fast formation of DCAA at pH 8 during the first hour of chloramine contact time (i.e., approximately 50% of the 72-hr DCAA yield) under the UFC protocol (Figure 1.2). Duirk (2003) reported gradually increasing

DXAA formation at longer reaction times; however, no detailed early kinetic data (before 24 hr) was reported in his study (data not shown). The limited data available in the literature about HAA formation kinetics during chloramination clearly indicated the need for further systematic studies, which became another objective of this study. The correct information about HAA formation kinetics is also critical for understanding the formation and speciation of HAAs and for developing models (mechanistic or empirical) to predict HAA formation during chloramination.

Important HAA Formation Factors

There are several factors that are important for HAA formation during chloramination, including DOM characteristics, pH, bromide concentration, temperature, chlorine to ammonia ratio, and the ammonia and chlorine addition point(s), and the extent of their mixing.

DOM Characteristics

DOM, which constitutes over 80% of the natural organic matter (NOM) in fresh waters, is a heterogeneous mixture of various organic molecules originating from aquagenic (i.e., biota in a water body) and pedogenic (i.e., soil and terrestrial vegetation) sources (Thurman 1985, Hayes et al. 1989, Glaze 1990, MacCarthy et al. 1990, Leenheer and Croué 2003). The physicochemical composition of DOM is influenced by such biogeochemical processes occurring in aquatic environments as fixation of carbon by algae and aquatic plants, transformation and biodegradation of organic materials, partitioning between solid and liquid phases, photo degradation, and oxidation (Aiken and Cotsaris 1995). In addition, runoff and storm events flushing organic matter from soil and plant debris, diffusion from sediments, and living or decayed vegetation are also major contributors to the organic matter matrix in natural waters (Krasner et al. 1996a). The presence of DOM leads to various problems in drinking water treatment processes and distribution systems, including increasing chemical (e.g., coagulant and oxidant) demands, reducing effectiveness of some treatment processes (e.g., adsorption and membranes), serving as precursor for DBP formation, providing substrate for biological re-growth in distribution systems, binding regulated metals and organic chemicals and transporting them, and creating taste and odor problems. As a result, although DOM itself is not regulated, water utilities usually try to maximize the removal of DOM at an affordable cost. Since DOM serves as precursors of DBPs, its characteristics play an important role in the formation and speciation of DBPs, as widely documented for chlorination. Therefore, DOM characteristics may also influence DBP formation during chloramination. As a result, investigation of DOM characteristics is important for the understanding of HAA formation mechanisms during chloramination. However, there is limited information in the literature on this topic.

Specific UV absorbance at 254 nm ($SUVA_{254}$, L-m/mg), which is the ratio of UV absorbance at 254 nm divided by the DOC (in mg-C/L) of a water sample, then multiplied by 100, is used as an indicator of the aromaticity of DOM and has been used to predict DBP formation in chlorination (Reckhow et al. 1990, Najm et al. 1994, Krasner et al. 1996a, Teng and Veenstra 1996, Croué et al. 2000, Kitis et al. 2002, Liang and Singer 2003). Differential spectroscopy at 272 nm has also used to probe the correlation of DOM nature and reactivity of DBP formation (Korshin et al. 1997b, Korshin et al. 1997a, Li et al. 2000, Duirk 2003). In addition, $SUVA_{254}$ measurements are included in the USEPA D/DBPR, as part of one of the alternative compliance criteria from the treatment technique (i.e., enhanced coagulation) requirements (USEPA 1998). Usually, aquagenic DOM contains mainly aliphatic organic substances, is low in phenolic and aromatic constituents, and has low $SUVA_{254}$ values (i.e., ≤ 3 L/mg-m). On the other hand, pedogenic DOM, which originates from soil and terrestrial vegetation, contains more aromatic moieties and lignin, and has relatively high $SUVA_{254}$ values (i.e., ≥4 L/mg-m). Since DOM is a heterogeneous complex, its fractionation into more homogenous components has been the most common approach for facilitating the interpretation of its characterization and reactivity studies. DOM has been frequently fractionated according to hydrophobicity and/or molecular weight (MW).

Resin adsorption chromatography (RAC) has been used to fractionate DOM into different fractions based on its hydrophobicity. This technique was initially developed by Thurman and Malcolm (1981) and Leenheer (1981). A modified version of the original resin fractionation method was later presented (Malcolm and MacCarthy 1992, Aiken et al. 1992). RAC is a chromatographic separation process that isolates and fractionates DOM into fractions with different structural and functional properties. Fractionation of DOM by RAC involves XAD-type resins, non-ionic macroporous materials, which concentrate and fractionate DOM into operationally defined fractions depending on their affinities to a given resin. RAC has been frequently used to obtain DOM with specific characteristics (Thurman and Malcolm 1981, Leenheer 1981, Malcolm and McCarthy 1992, Town and Powell 1993, Tadanier et al. 1999, Croué et al. 2003, Chow et al. 2006). While this method can provide DOM that accommodate specific research needs and is also relatively free from confounding effects of inorganic substances, there are uncertainties over possible chemical alteration of DOM due to harsh pH changes during the isolation process, irreversible adsorption to the resins, contamination from resin bleeding, and size-exclusion effects (Malcolm 1991, Malcolm and McCarthy 1992, Town and Powell 1993). Furthermore, the fractionation of DOM is largely dependent on the operational conditions employed. The column capacity factor (k'), an important parameter that determines hydrophobic/hydrophilic distribution of DOM, may significantly vary among studies, however it is not always reported. The large variation (or lack of reporting) of k' in the literature prevents direct comparison of fractionation data and fraction related results (e.g., characterization and reactivity) obtained from different experiments. During the RAC procedure, the pH of DOM solution is first adjusted to 2 and then it is applied onto a macroporous and nonioinic resin (XAD-8) (Aiken et al. 1979, Thurman and Malcolm 1981). The eluent from the XAD-8 column is applied to an XAD-4 column. The fractionation process results in three operationally defined fractions. The fractions desorbed from the XAD-8 and XAD-4 columns are hydrophobic (HPO) and transphilic (TPH) fractions, respectively, and the eluent from XAD-4 column (i.e., the DOM components that were not removed by any column) is called the hydrophilic (HPI) fraction. The transphilic term is used to describe fractions of an intermediate polarity between hydrophobic and hydrophilic fractions. Each fraction can also be further separated into its corresponding acid, base, or neutral fractions. More recently, Leenheer et al. (2000) expanded the RAC fractionation method to recover the colloidal DOM fraction in natural waters.

A few studies provided information about HAA reactivity of DOM fractions during chloramination. Duirk (2003) found a positive correlation between $SUVA_{280}$ (i.e., UV absorbance at 280 nm divided by the DOC) and DCAA formation coefficients for three isolated fractions in three natural waters, but not for a commercial humic acid solution. Speitel et al. (2004) reported that waters after conventional treatment processes always had lower DXAA reactivity (on a per organic carbon basis) than the source waters, which indicated that HAA precursors were hydrophobic in character, because hydrophobic fractions were more amenable to coagulation than hydrophilic fractions (Collins et al. 1986, Singer and Harrington 1993). Hua and Reckhow

(2005b) fractionated three surface waters, and the chloramination results showed HPO and TPH fractions were more reactive in forming DXAA than HPI fractions in a water with high $SUVA_{254}$ (4.4 L/mg-m); however, in two low $SUVA_{254}$ waters (2.8 and 1.6 L/mg-m), the DXAA reactivity was similar for all fractions at pH 7. In contrast to these reports, Hwang et al. (2000) studied Colorado River water, a very low $SUVA_{254}$ water (1–1.6 L/mg-m), and showed that the yields of DXAA from hydrophilic acid and neutral fractions were almost 3 times more than that from the hydrophobic acid fraction. Furthermore, the hydrophilic fraction showed the highest reactivity toward THM formation, and the DXAA yield was higher than the TXAA yield during chlorination. Hwang et al. (2000) explained this difference using Reckhow and Singer's (1985) model, which suggested high content of ketoacids (R'-CO-CH_2-COOH) in this low SUVA water favored formation of DXAA.

Although positive correlations between HAA yields and $SUVA_{254}$ of DOM fractions have been reported for both chlorination and chloramination, the effect of $SUVA_{254}$ is much less pronounced during chloramination. In general, better correlations were obtained for the DOM fractions of the same source water. The correlations usually weaken when the reactivity of DOM or DOM fractions from various sources were compared in a single correlation (Symons et al. 1998, Wu et al. 2003, Speitel et al. 2004). The weak relationships probably result from the heterogeneity of DOM characteristics, and $SUVA_{254}$ is not always sufficient or sensitive enough to distinguish among the reactivity of DOM from various sources toward the HAA formation.

It has been reported that MW was not a good indicator for HAA formation during chlorination (Teng and Veenstra 1996, Sinha et al. 1997, Kitis et al. 2002). There is limited information in the literature about the effect of DOM MW on the HAA formation during chloramination. Recently, Hua and Reckhow (2005b) fractionated three surface waters and reported that no significant trend between MW for THM and HAA formation during chloramination, and the formation was very similar for all fractions of a low $SUVA_{254}$ water ($SUVA_{254}$ = 1.6 L/mg-m).

pH

The residual NH_2Cl concentration and TOX and HAA formation depend strongly on pH. Speitel et al. (2004) reported that pH is the most significant factor in DXAA formation under typical drinking water chloramination conditions, within the pH range of 7 to 9 and Br^-/DOC ratios less than 100 μg/mg. However, it has been shown that the impact of pH on DXAA formation is not proportional to pH (Duirk 2003). The DXAA formation decreased significantly as the pH increases during chloramination. Moreover, in spite of a very low degree of THM formation, it was reported that the formation of THMs also decreased with increasing pH during chloramination (Diehl et al. 2000).

Bromide

Similar to chlorination, the presence of bromide increases HAA formation during chloramination, shifting the speciation to more brominated species, but to a lesser extent, as compared to chlorination. A high Br^-/DOC ratio in drinking water is undesirable because it increases the HAA mass concentration because the bromine-substituted DBP species have higher molecular weights than the chlorinated DBPs. Moreover, one of the bromine-substituted HAA species is believed to pose a greater health concern than its chlorinated counterpart (Bull et al. 2001). Under typical drinking water conditions with Br^-/DOC <100 μg/mg, bromide did not increase HAA formation significantly during chloramination (Cowman and Singer 1996, Diehl et al. 2000, Speitel et al. 2004). In general, higher Br^-/DOC ratio and lower pH lead to a higher degree of bromine incorporation into HAAs (Diehl et al. 2000, Duirk 2003, Qi et al. 2004, Speitel et al. 2004). Brominated species always exhibit faster formation kinetics and higher molar yields than those of chlorinated species. The effect of bromide on HAA formation during chloramination decreases with increasing pH.

The effect of bromide depends also on DOM characteristics. Duirk (2003) developed a model by using a single DOM specific rate constant ($k_{doc,Br}$) to characterize the reaction between active bromine (Br in +1 oxidation state) and DOM during chloramination. The specific rate constant was found to correlate quite well with $SUVA_{280}$, indicating that active bromine reacted with the aromatic portion of the DOM structure. Wu et al. (2003) compared correlations of $SUVA_{254}$ with the bromine incorporation factor (BIF) in seven humic substances isolated from surface waters, soils and sediments during chloramination, and they observed no correlation ($r = 0.08$). Examining the reactivity of all DOMs from very different sources in a single correlation was probably the main reason for the observed poor relationship. The correlation significantly improved ($r = 0.96$) when four surface water sources were compared, omitting the DOM sources from soils and sediments.

Cl_2/N Ratio

The Cl_2/N ratio is another important contributor to DXAA formation, in addition to pH and bromide, under typical drinking water chloramination conditions (Speitel et al. 2004). In general, higher Cl_2/N ratio favors DCAA formation because of the higher HOCl concentration released from NH_2Cl hydrolysis (Qi et al. 2004). It has been shown that lower Cl_2/N ratios can reduce HAA formation (Diehl et al. 2000, Qi et al. 2004). However, low Cl_2/N ratios are not desirable in practice, because of the higher potential of nitrifying bacteria growth in distribution systems.

Temperature

The formation of HAAs increases with temperature due to the faster kinetics. Speitel et al. (2004) reported that DXAA formation increased moderately in the range of 21 to 29°C. Hua and Reckhow (2005a) observed a 20% and 37% increase of HAA formation when the temperature increased from 5 to 15 and 30°C, respectively. The effect of temperature was more significant as the pH decreased and the Cl_2/N ratio increased. However, Baribeau et al. (2005) noted that both higher temperatures and lack of NH_2Cl residuals resulted in biodegradation of DCAA significantly in distribution systems. In the absence of a NH_2Cl residual, 60% of DXAAs degraded in warm waters (17–22°C), but no degradation was observed under colder temperatures (12–14°C). Higher temperatures can accelerate NH_2Cl residual depletion and DCAA degradation. Therefore, increased HAA formation kinetics with high temperature may be counter-balanced with an increase in biological activity in distributions systems if the residual disinfectant is not maintained. Observing a decreasing trend of the HAA concentrations with increasing residence times or locations with lower concentrations at warm temperatures will be an indication of biological activity in distribution systems.

Chloramine Formation in Practical Applications

The formation of NH_2Cl can be accomplished by adding ammonia and chlorine. The mixing order of these two chemicals can lead to significantly different DBP formation patterns among utilities practicing chloramination. If chlorine (Cl_2) is added first, the lag time until ammonia addition allows DOM to react with HOCl, significantly increasing DBP (both HAA and THM) formation prior to NH_2Cl formation. Therefore, if utilities are not concerned with the CT credit from chlorine, which is the product of the disinfectant concentration (C, in milligrams per liter) and disinfectant contact time (T, in minutes), minimizing the lag time between ammonia and chlorine addition or addition of ammonia prior to chlorine can provide a practical solution for decreasing THM and HAA formation as long as an alternative primary disinfectant (i.e., ozone, chlorine dioxide, UV) is used.

The mixing condition at the chlorine and ammonia addition point may play an additional role in DBP formation. Symons et al. (1998) compared DBP formation for chloramination (with 30-second prechlorination) under different mixing conditions. They performed experiments at three mixing intensity levels (i.e., mean velocity gradient (G)). The high ($G = 1000\ sec^{-1}$) and medium ($G = 500\ sec^{-1}$) mixing conditions produced less THMs than low mixing intensity ($G = 60\ sec^{-1}$), while the effect on HAA formation was not significant. The effect of mixing condition may be more significant when shorter prechlorination time is employed. If ammonia and chlorine are added in water approximately at the same location, the formation of THMs should be minimal, because chloramination does not produce significant amounts of THM. Observing high THM concentrations under such circumstances would be an indication of mixing problems.

Overall, it should be evident from this review that a combined evaluation and optimization of several factors reviewed in this section (i.e., DOM characteristics, reaction time, pH, bromide, Cl_2/N, temperature, chloramine application techniques) is needed to comply with the stringent requirements of the Stage 2 D/DBPR using chloramination.

SUMMARY OF IMPORTANT FINDINGS FROM THE LITERATURE

A summary of important findings from the presented literature review is as follows:

- THM formation is almost completely suppressed using preformed NH_2Cl.
- DXAA is the primary class of DBPs formed during chloramination using preformed or *in-situ* formed chloramines. Formation of MXAAs and TXAAs are usually negligible. This speciation pattern differs from chlorination, for which the amounts of both DXAAs and TXAAs formed can be significant.
- It has been hypothesized in the literature that the small amount of HOCl resulting from NH_2Cl hydrolysis is the major reactant with DOM and forms DXAAs during chloramination. The significance of NH_2Cl and $NHCl_2$ as possible halogenation agents reacting with DOM to form DXAAs is not known.
- The impact of DOM characteristics as a precursor of DXAAs during chloramination is not widely documented. The limited data available in the literature suggest that hydrophobic fractions of DOM, in general, result in higher DXAA formation than other fractions (i.e., transphilic and hydrophilic). However, in some cases, very high reactivity of hydrophilic DOM components toward HAA formation has also been reported.

- There are limited but mostly inconsistent data in the literature regarding the HAA formation kinetics during chloramination. The causes of these inconsistencies were unknown.
- DXAA formation during chloramination depends on many factors, including pH, bromide, Cl_2/N ratio, and temperature. DXAA formation decreases with increasing pH. This is different from the effect of pH on DXAAs during chlorination, where DXAA formation remains relatively constant with changes in pH. The presence of bromide can increase DXAA formation during chloramination, shifting the speciation to more brominated species, but to a much lesser extent than in chlorination. Limited information indicated that the effect of bromide may depend on the DOM characteristics. In general, higher Cl_2/N ratios favor DXAA formation. However, lower Cl_2/N ratios are not desirable due to increased potential for nitrification in distribution systems. DXAA formation increases with temperature, while increasing temperature also increases its biodegradation rates in a distribution system with low to no residual.
- The lag time between chlorine prior to ammonia addition during chloramination significantly increases DXAA formation and results in TXAA formation. For utilities not concerned about the CT credit from chlorination, reducing this lag time or addition of ammonia prior to chlorine can significantly decrease HAA formation as long as an alternative predisinfectant (i.e., ozone, chlorine dioxide, UV) is used.
- TOX formation, representing the overall DBP formation, is significantly lower during chloramination than during chlorination. However, the fraction of unidentified DBPs during chloramination is higher than that during chlorination.

RESEARCH OBJECTIVES

It is evident from the literature review presented in this section that there is still a lot to learn about the formation and speciation of HAAs during chloramination. This is especially important considering the significant increase in the use of chloramines in the drinking water treatment industry in the US. The research objectives for this project are grouped into three areas:

1. *The Pathways of HAA Formation:* Today, the widely accepted hypothesis for the formation of HAAs during chloramination is that the small amount of HOCl resulting from NH_2Cl hydrolysis is the major reactant forming DXAAs during chloramination. However, NH_2Cl hydrolysis is a very slow reaction and it was calculated that HOCl only contributes less than 30% of the TOX formation during chloramination. The significance of the reactions of NH_2Cl and $NHCl_2$ with DOM, the other two halogenation agents that can be present during chloramination, is not known. Therefore, the first objective in this research was to investigate the importance of possible reaction pathways in HAA formation.
2. *The Kinetics of HAA Formation:* There is limited and mostly inconsistent data in the literature regarding the HAA formation kinetics during chloramination, although some of these data were obtained under comparable experimental conditions. Therefore, the second objective in this project was to investigate the potential causes of this variability.
3. *The Roles of DOM Characteristics, pH and Bromide on HAA Formation:* It is not well understood how different components in a DOM mixture contribute to HAA formation

and speciation during chloramination, and how HAA formation is related to some characteristics of DOM. Therefore, the third objective in this research was to systematically investigate the formation and speciation of HAAs from different DOM components in natural waters. The roles of pH and bromide in HAA formation and speciation were also examined due to the practical significance of these parameters at typical water treatment conditions. Furthermore, kinetics experiments were conducted to examine and compare HAA formation kinetics from different DOM fractions at different pH and bromide conditions.

A systematic investigation of these issues is important to: (i) gain a fundamental insight into the formation of HAAs, especially DXAAs, the dominant form of HAAs during chloramination; (ii) develop models for predicting HAA formation and speciation; and (iii) develop strategies for controlling HAA formation during drinking water treatment, thus complying with the USEPA's increasingly stringent D/DBP Rule requirements.

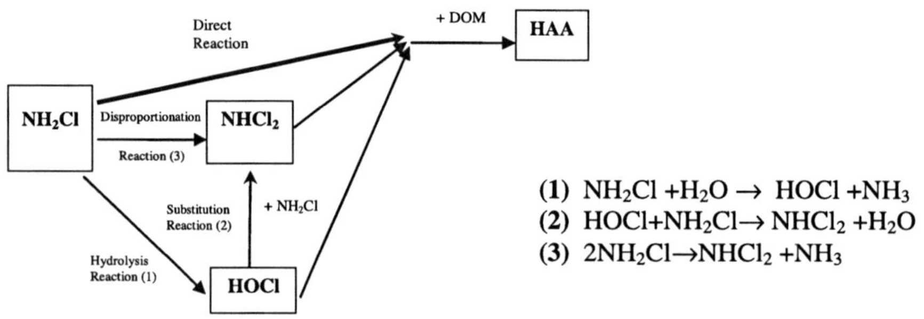

Figure 1.1 Conceptual illustration of HAA formation pathways in the absence of bromide during chloramination

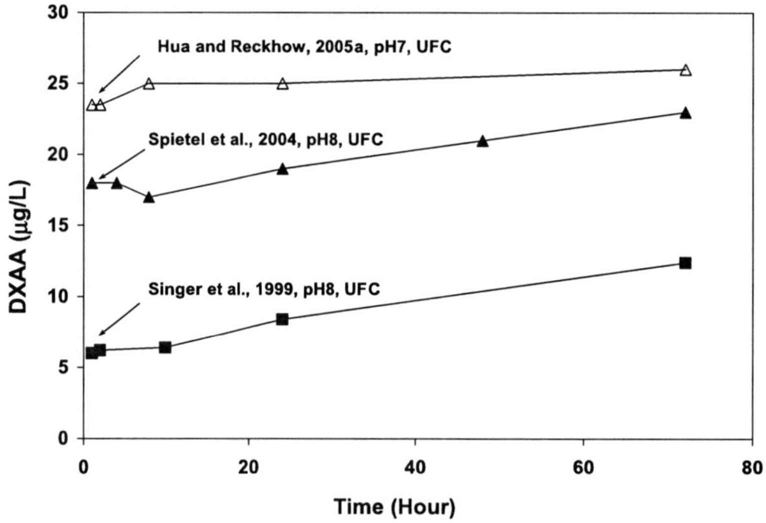

Figure 1.2 DXAA formation kinetics using preformed chloramine, reported by different researchers

CHAPTER 2
MATERIALS AND METHODS

SOURCE WATERS

Surface water DOMs were collected from the influents of three drinking water treatment plants (WTPs) in South Carolina: Greenville (North Saluda and the Table Rock Reservoirs), Myrtle Beach (Intercostal Water Way), and Spartanburg (Lake Bowens) using a reverse osmosis (RO) system. The names in the parentheses are the names of the water sources for the treatment plants.

REVERSE OSMOSIS SYSTEM

Figure 2.1 schematically depicts the field-scale RO system used for DOM isolation in this project. The system was constructed with stainless steel components to the greatest extent possible. The characteristics of the system are summarized in Table 2.1. During field operation, source water was obtained directly from the influent of each drinking water treatment plant before any chemical addition. A series of pretreatment steps was initially performed before applying the water to the RO membrane. Pretreatment included: 1) a series of pre-filters to remove larger particles; 2) a (sodium-form) cation exchange softener to remove multivalent cations (mainly Ca^{++} and Mg^{++}) and minimize precipitation (scaling) on the membrane surface; and 3) a final cartridge filter (0.45-μm). After pretreatment, filtered and softened source water was collected in a 55-gallon high-density polyethylene tank (not shown in Figure 2.1), which was sampled for mass balance calculations. The water volume in the tank was also recorded. Then the water was emptied into a 55-gallon stainless-steel drum that served as a feed reservoir. DOM feed solution was pumped to a high-pressure stainless-steel multi-stage centrifugal pump using a stainless steel centrifugal transfer pump. The high-pressure pump provided the necessary flow (about 4 gpm or 15 L/min) and pressure (up to 210 psig) for efficient membrane operation. Retentate was routed back to the isolate reservoir while permeate was collected in 20-L fluorinated polyethylene jerricans, which were sampled for mass balance calculations, and the volumes were also recorded. Some of the permeate water was saved for system cleaning. The system was designed for a feed flow of about 4 gpm, a permeate flow of about 0.5 to 0.7 gpm and a retentate flow of about 3.3 to 3.5 gpm, or about 5 to 7 times the permeate flow. This ratio was recommended to provide adequate cross-flow velocity to minimize membrane fouling. Pressure gauges (P) were installed to monitor pressure in the system. A back-pressure needle valve was used to control the operating pressure in the membrane module. Most field operations employed a transmembrane pressure of about 550 kPA (80 psig), which has been found to minimize membrane fouling during our recent work (Kilduff et al. 2004).

An isolation run was concluded by stopping the flow of pretreated source water to the feed reservoir, and then allowing the volume in the feed reservoir to decrease to approximately 20 L. At the end of the process, the water in the void volume of the system (e.g., membrane housing, pumps and lines) was emptied. After recovering the concentrated isolate, the system was flushed and cleaned by recirculating 40 L of permeate (or distilled deionized water [DDW] in the laboratory) adjusted to pH 10 (using NaOH) for about 20 minutes. After chemical cleaning, the system was again flushed with permeate water (or DDW in the laboratory) until a neutral pH (about 7) was obtained. Overall mass recovery of a particular parameter (e.g., DOC) was calculated from

Table 2.1
Characteristics and specifications of the field-scale RO system

Parameter	Specification
Membrane*	
Size	AK4040TF (3.9 inch × 40 inch)
Module type	Spiral-wound cross-flow
Material	Aromatic polyamide thin film composite
Total surface area	89 ft^2
Maximum operating pressure	110 psig
Salt rejection @ 225 psig (2000 ppm NaCl)	99% (typical)
Membrane Feed Water	
Silt density index (15 min)	less than 5
pH range	3–11
Temperature range	10–30°C
Maximum chlorine content	less than 0.1 mg/L
Maximum flow	6 gpm
Maximum turbidity	less than 1 NTU

* Osmonics, Inc.

concentration measurements and the volumes of recovered solutions (including isolated concentrate, permeates, NaOH wash, water flush, and analytical samples):

$$\%\ \text{Total Mass Recovery} = \frac{V_c C_c + \sum V_{permeate} C_{permeate} + V_{NaOH} C_{NaOH} + V_{flush} C_{flush}}{V_{source} C_{source}}$$

where, V_c and C_c are the volume and concentration of isolate (~20 L), the sum of $V_{permeate} \times C_{permeate}$ is the mass (mg) of a parameter (e.g., DOC) in all permeates collected during the isolation; $V_{NaOH} \times C_{NaOH}$ is the mass (mg) of the parameter present in the NaOH cleaning solution; $V_{flush} \times C_{flush}$ is the mass of (mg) of the parameter present in the flush water (either DDW or permeate); and $V_{source} \times C_{source}$ is the mass of (mg) of the parameter in the pretreated source water feed that is applied to the membrane. Where possible, sample volumes were measured gravimetrically using an electronic balance.

DOM FRACTIONATION

The DOM concentrate obtained with the RO system was fractionated using a procedure adapted from Leenheer et al. (2000). Figure 2.2 shows a schematic of the fractionation procedure. There are two parameters that are important in designing fractionation experiments using this procedure as well as for comparing results from different studies and source waters: k' and C_o (DOM concentration). The k' is directly proportional to the volume of water processed through the resin bed (Leenheer 1981). As k' increases, available adsorption sites on the resin are occupied, resulting in the breakthrough of higher UV-absorbing DOM components (Kitis et al. 2001). The effect of C_o on DOM fractionation for a constant k' is not well documented. Therefore, the resin fractionation procedure is operationally defined. In this project, k' of 15 (i.e., 20 L of water applied for the column system used in this study for the fractionation) and C_0 of 150 mg DOC/L were selected for the fractionation. These two parameters were kept constant for all waters.

During fractionation, DOM concentrate was initially applied to a pre-cleaned XAD-8 resin column at pH 2. The fraction adsorbed by the resin and subsequently back-eluted from the column using acetonitrile is designated as HPO. After acetonitrile extraction, a 0.1 N NaOH solution is used to remove the remaining HPO fractions and to further clean the resin before the next use. The NaOH elution was included in the mass balance calculations but it was not used for the DBP experiments. The permeate from the XAD-8 column was then applied to a pre-cleaned XAD-4 column. The fraction adsorbed by the resin and subsequently back-eluted from the column using acetonitrile was designated as TPH. Similar to the procedure for the XAD-8 resin, XAD-4 resin was also further cleaned with NaOH elution. The fraction that did not adsorb on the resin was in the permeate of XAD-4 resin and it was designated as HPI. After acetonitrile extraction, the recovered HPO and TPH fractions were rotaevaporated to remove acetonitrile and to concentrate the fractions. They were then lyophilized using a freeze dryer and stored in powder form in a dessicator. The HPI fraction was kept in aqueous solution and stored in a refrigerator at 4°C around pH of 4.

During the resin fractionation most of the inorganic components in water ends up in the HPI fraction. The efficacy of dialysis to reduce the background inorganic concentration, especially the bromide level, was evaluated using the Greenville HPI fraction. Several dialysis bags (Spectra/Por Cellulose Ester Membranes) with molecular weight cutoff values of 100 were filled with the HPI fraction. The bags were placed in a rectangular glass container filled with 25 liters of DDW. The DDW in the container was replenished three times during the course of the experiment to maintain a high driving force (i.e., concentration gradient) for dialysis. The change in the organic carbon concentration of the DDW as a function of time was monitored. Finally, the DOC concentration of HPI fraction remaining in the bags at the end of dialysis was measured. Mass balance calculations were performed and the results are provided in Table 2.2. A good mass balance closure was obtained for the experiment (i.e., 107%). After 6 days of dialysis, it was found that a significant fraction of HPI components (i.e., ~80%) passed through the dialysis bags, indicating that the Greenville HPI fraction is composed of primarily small molecular weight organic molecules. Due to high organic carbon losses observed, it was decided not to use dialysis during the course of the project. In addition, there is no evidence in the literature indicating that the common background inorganic composition of natural waters, except pH, has a significant impact on the HAA formation and speciation. Therefore, reducing the background inorganic levels in the HPI fractions was not critical for performing chloramination experiments, as long as bromide concentration was carefully monitored.

CHLORAMINATION EXPERIMENTS

All samples were chloraminated according to the UFC protocol originally developed for chlorination (Summers et al., 1996), with the following modifications:

a. Experiments were conducted at three different pHs, not only at pH 8 as prescribed in the UFC protocol, in order to investigate the pH effect,
b. The total combined chlorine residual after 24 hr was maintained around 2 mg/L as Cl_2, instead of 1 mg/L free chlorine residual, and
c. Carbonate buffer was used instead of borate buffer.

Table 2.2
Dialysis of Greenville HPI fraction

	Time elapsed (hr)	DOC concentration (mg/L) In dialysis bag	DOC concentration (mg/L) In the DDW*	Volume (L)	DOC mass (mg) In dialysis bag	DOC mass (mg) Loss to DDW
HPI applied	0	136.1	0.2	0.52	70.8	
First dialysis solution	24	NA	0.8†	25		
	51	NA	1.7†	25		42.5‡
Second dialysis solution	82	NA	0.5†	25		11.5
Third dialysis solution	144	NA	0.1†	25		2.5
HPI after dialysis		37.2§		0.52	19.3	
Total HPI lost						56.5

NA: Not available.
* Distilled and deionized water used for dialysis.
† DOC concentration after correction with the background DOC of DDW used for dialysis.
‡ Total DOM mass lost at the end of 51 hours during the first dialysis.
§ DOC concentration of HPI remained in the bags at the end of dialysis.
Mass balance:

$$\frac{\text{DOC remaining after dialysis (19.3 mg) + DOC lost during dialysis (56.5 mg)}}{\text{DOC applied (70.8 mg)}} \times 100 = 107\%$$

These modifications in the UFC protocol represent typical conditions during chloramination in practice. According to the American Water Works Association disinfection committee survey, the mean and median for total chlorine residuals in medium and large sized chloraminated distribution systems were 2.4 and 2.5 mg/L, respectively (American Water Works Association Committee Report 2000). Although the average detention time in the distribution systems was not provided in this report, it is reasonable to use 2 mg/L total residual after 24 hr in the UFC tests for several reasons: 1) NH_2Cl is not very reactive and the proposed 2 mg/L residual is close to the average concentration in the chloramination systems, 2) the survey conducted for the development of the UFC protocol indicated a 1.3-day average residence time in the distribution systems for chlorinating utilities, and 3) other researchers have used similar conditions in different chloramination studies (e.g., Singer et al. 1999).

Before each chloramination experiment, a fresh NH_2Cl stock solution was prepared by mixing sodium hypochlorite (5–6% available free chlorine) and ammonium sulfate solutions in DDW at a Cl_2/N mass ratio of 3.5:1 at pH 9.0. Preliminary experiments were conducted for each sample to determine the ratio of NH_2Cl/DOC that would provide a total chlorine residual of about 2.0 mg/L for chloramination after one day and a detectable residual after 72 hr of contact time. Kinetic experiments were performed for 0.5, 1, 2, 4, 8, 24, 48, and 72 hr. At the end of each target contact time, samples were collected for pH, and free (if there was any) and combined chlorine residual. Then, samples were immediately quenched with a stoichiometric amount of sodium sulfite (Na_2SO_3) and analyzed for THMs and HAAs.

For $NHCl_2$ experiments, a preformed $NHCl_2$ stock solution was prepared by rapidly adjusting the pH of the NH_2Cl stock solution to 3.5–4 with the addition of 1M of a hydrochloric acid (HCl) solution. The solution was aged for 4 hr. The dose of $NHCl_2$ was chosen based on its expected occurrence during chloramination practices. In limited experiments, chlorination was performed at similar conditions to that of chloramination; the only difference was that the target residual of chlorine was 1±0.4 mg/L as Cl_2 after 24 hr.

ANALYTICAL METHODS

A number of analytical methods (USEPA or *Standard Methods*) were used to measure various parameters during the course of this study. In this section, the details of HAA and THM measurement methods will be described. The details of other methods can be found in USEPA manuals and *Standard Methods* (APHA et al. 1998). The parameters, measurement methods, the equipment used, and minimum reporting levels or accuracy are listed in Table 2.3. DDW that was used to dilute samples and prepare standards or reconstituted waters had a DOC of less than 0.2 mg/L and ionic strength of less than 0.006 mM.

Measurements of HAA and THM

In this project, the calibration standards were prepared in the filtered natural water samples to be used for the chloramination experiments. Therefore, it was possible to account for any background matrix effect on the analytical methods. Samples were filtered using pre-washed 0.45 μm Gelman Supor filters. In order to minimize the impact of temporal variability on the results, as a general practice, standards were prepared at the same time with the samples for both THM and HAA measurements. Both samples and calibration standards were treated in the same manner.

HAA

For HAA measurements, the analytical method developed by Singer et al. (2002) using diazomethane derivatization based on Standard Method 6251B was closely followed. Since the method and step-by-step analytical procedure was described in the AwwaRF report entitled "Relative Dominance of HAAs and THMs in Treated Drinking Water," they will not be repeated here. An Agilent 6890 gas chromatograph equipped with an electron capture detector and an autosampler was used for the measurements.

Detection limit (DL) is the lowest concentration of an analyte for which it can be reported with 99-percent confidence that the concentration is greater than zero (Glaser et al. 1981). DL is obtained using the following equation:

$$DL = S * t_{(n-1,\ 1-\alpha)}$$

where

S = standard deviation of the replicate analyses

$t_{(n-1,\ 1-\alpha)}$ = student-t value for the 99% confidence level with n–1 degree of freedom, (e.g., $t_{(6,0.99)} = 3.143$ for 7 replicates)

n = number of replicates

$\alpha = 0.01$, confidence level is $1-\alpha = 0.99$

Minimum reporting level (MRL) is the threshold concentration of an analyte that a laboratory can expect to accurately quantify in an unknown sample. The MRL is usually established at a

Table 2.3
Analytical methods and minimum reporting levels

Parameter	Unit	Method	Equipment	Minimum reporting level or accuracy[*]
DOC[†]	(mg/L)	SM[‡] 5310B	TOC-V CHS & TNM-1, Shimadzu Corp., Japan	0.1
Total Nitrogen	(mg/L)	High Temperature Combustion	TOC-V CHS & TNM-1, Shimadzu Corp., Japan	0.1
UV Absorbance[§]	(cm^{-1})	SM 5910	DU 640, Beckman Inst. Inc., USA	±0.005[**]
Total Alkalinity	(mg/L as $CaCO_3$)	SM 2320	NA[††]	2–4
Ca & Mg Hardness	(mg/L as $CaCO_3$)	SM 3500 (Ca-B and Mg-B)	ICP-AE, Spectra CIROS, USA	Ca: 0.25; Mg: 0.025
Br^-, NO_3^-, NO_2^-	(μg/L)	USEPA Method 300	ED 40, Dionex Corp., USA	Br^-: 25; NO_3^- and NO_2^-: 50
Conductivity[‡‡]	(μS/cm)	SM 2510	M90, Corning Corp., USA	±0.5–1%[§§]
NH_3	(mg/L)	SM 4500-NH_3 F	NH_3 Selective Electrode	0.1
pH		SM 4500-H^+	420A, Orion Corp., USA	±0.01[***]
THMs[†††]	(μg/L)	USEPA Method 551.1	6890 GC-ECD, Agilent, USA	1
HAA_9[‡‡‡]	(μg/L)	SM 6251 B[§§§]	6890 GC- ECD, Agilent, USA	1
Residual Free/Combined Chlorine	(mg/L)	SM 4500-Cl F	NA	0.1–0.15

* As reported by the manufacturer.
† Reagent grade potassium hydrogen phthalate was used to prepare external standards. Precision ranged from 0.05 to 0.15 mg/L.
‡ SM: *Standard Methods* (APHA, AWWA, and WEF 1998).
§ Measured at wavelengths of 254, 272 and/or 280 nm using a 1- or 5-cm cell.
** Photometric accuracy (absorbance units).
†† NA: Not applicable.
‡‡ NaCl standards were used to correlate conductivity readings to ionic strength.
§§ Relative accuracy.
*** Accuracy (pH units).
††† THMs: Trihalomethanes. Methyl-tert butyl ether (MTBE) solvent extraction and a gas chromatograph (GC) with an electron capture detector (ECD) analysis.
‡‡‡ HAA_9: All nine haloacetic acid species.
§§§ Diazomethane derivatization, MTBE solvent extraction and GC-ECD analysis was used by closely following the detailed procedures described in Chapter 2 and in Appendix A of the AwwaRF report entitled "Relative Dominance of HAAs and THMs in Treated Drinking Water" by Singer et al. (2002). The results of the method development for HAA_9 in this project are summarized in Appendix A of this report.

Table 2.4
Detection limits of HAA species in Greenville water

Analyte	Spiked concentration (μg/L)	Mean measured concentration (μg/L)	SD (μg/L)	RSD (%)	DL (μg/L)	MRL (μg/L)
CAA	1.0	1.7	0.1	8	0.3	0.9
BAA	1.0	1.0	0.04	4	0.1	0.3
DCAA	1.0	0.8	0.05	7	0.2	0.6
BCAA	1.0	1.0	0.01	1	0.03	0.1
TCAA	1.0	1.0	0.01	1	0.03	0.1
DBAA	1.0	1.1	0.02	2	0.1	0.3
BDCAA	1.0	1.4	0.03	2	0.1	0.3
DBCAA	1.0	1.5	0.04	3	0.1	0.3
TBAA	1.0	1.6	0.07	4	0.2	0.6

concentration that is at least 3 times the DL or at least five times higher than the background noise. In practice, this is the lowest point on the calibration curve that can be practically obtained.

To determine the detection limits of all nine HAA species, a stock solution containing 1 μg/L of each HAA species was prepared in water samples initially collected from the influents of Greenville and Myrtle Beach water treatment plants. Because the results were very similar, only results from Greenville water are presented below. The stock solution was split into seven sub-samples. Each sub-sample was derivatized and extracted separately, and analyzed. Typical calibration curves were obtained for HAAs by collecting at least 6 points, with an R square higher than 0.99. The results of DL experiments and the determined MRLs are provided for Greenville water in Table 2.4. The percent relative standard deviations (RSDs) were less than 10%. The MRLs were set at 1 μg/L for each HAA specie.

Spike recovery experiments were performed to verify that the employed analytical method would be applicable to other water matrices. This issue was examined by analyzing spike recoveries of the HAA species in the tap water of our laboratory. Two sets of experiments were performed on two different days about a month apart. Each day, four tap water samples were collected. Duplicate samples were analyzed for HAA, and the other two samples were spiked with a cocktail containing 10 μg/L of each HAA specie before analysis.

The spike recovery was calculated using the following equation:

$$\text{Spike Recovery} = (M/(S+B)) * 100\%$$

where

M = measured concentration with spike (μg/L)
S = concentration of spike added (fortified concentration [μg/L])
B = background concentration (μg/L)

Table 2.5
Spike recoveries of HAA species in Clemson tap water

Analyte	Average background concentration (μg/L)	Fortified concentration (μg/L)	Mean measured concentration (μg/L)	Mean recovery (%)
CAA	3.3	10.0	10.8	81
BAA	<1	10.0	10.5	105
DCAA	22.3	10.0	31.5	97
BCAA	4.6	10.0	14.3	98
TCAA	14.2	10.0	24.6	101
DBAA	<1	10.0	10.6	106
BDCAA	6.5	10.0	17.1	104
DBCAA	1.5	10.0	11.8	103
TBAA	<1	10.0	11.6	116

Table 2.6
Detection limits of THM species in Greenville water

Analyte	Spiked concentration (μg/L)	Mean measured concentration (μg/L)	SD (μg/L)	RSD (%)	DL (μg/L)	MRL (μg/L)
TCM	1.0	1.1	0.04	3	0.1	0.3
DCBM	1.0	1.0	0.02	2	0.1	0.3
DBCM	1.0	1.0	0.02	2	0.1	0.3
TBM	1.0	1.1	0.05	5	0.2	0.6

Table 2.7
Spike recoveries of THM species in Clemson tap water

Analyte	Average background concentration (μg/L)	Fortified concentration (μg/L)	Mean measured concentration (μg/L)	Mean recovery (%)
TCM	29.1	10.0	38.0	97
DCBM	12.4	10.0	22.8	102
DBCM	3.0	10.0	13.9	106
TBM	<1	10.0	11.5	115

The results are shown in Table 2.5. The acceptable range for spike recoveries is within 80 to 120 percent. As the recoveries for HAA species were within this range, the method was found to be applicable for various natural water samples to be analyzed in this project.

THM

The approach used to determine DLs and MRLs for HAAs was also applied to THMs. USEPA method 551.1 for the liquid-liquid extraction and GC measurement of THMs was employed with minor modifications (USEPA 1996). Agilent 5890 GC with an electron capture detector, an autosampler, and a HP Chemstation software was used for the measurements. The DL was established at 1 μg/L for each THM specie in Greenville water (Table 2.6). Similar results were also obtained in Myrtle Beach water. The percent RSD's were below 10%. The MRLs were set at 1 μg/L for each THM specie. Typical calibration curves were obtained for THMs by collecting at least 6 points, with an R square higher than 0.99. As for HAAs, spike recovery experiments were performed for THMs. These experiments were performed on two separate days one month apart, as conducted for the HAAs. The results are presented in Table 2.7. The recoveries for THMs were all between 80 and 120 percent, indicating the analytical method was applicable for various samples that were tested during this project.

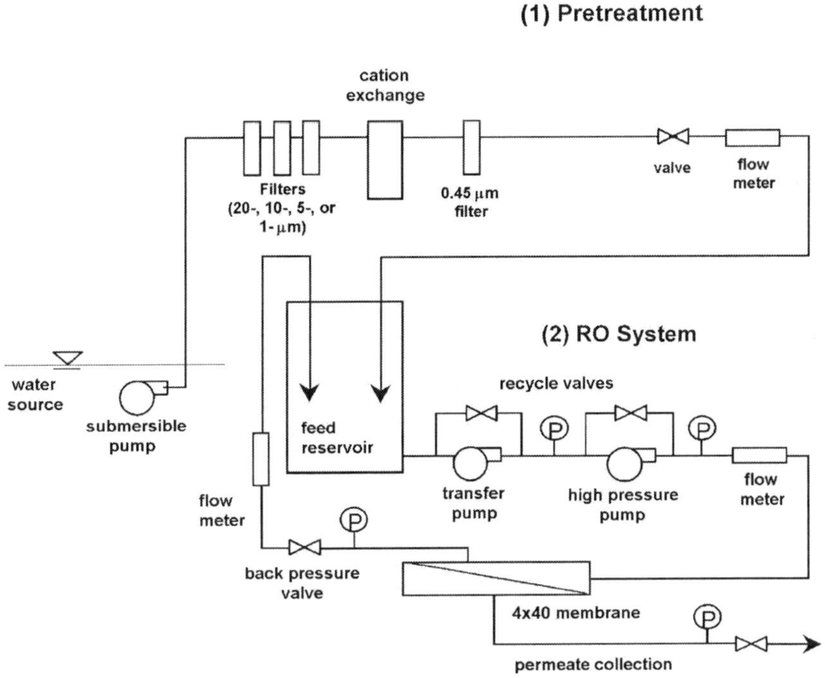

Figure 2.1 Schematic of field-scale RO system

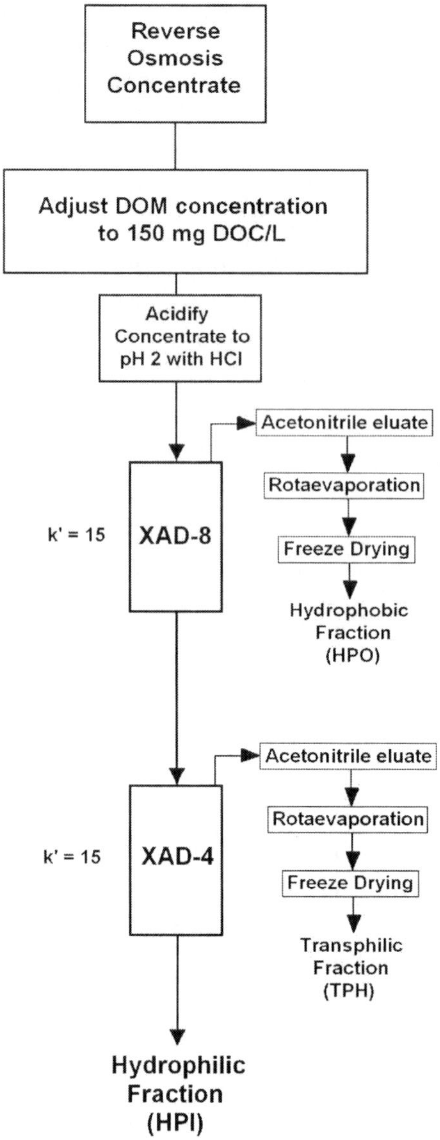

Figure 2.2 Schematic of DOM fractionation procedure

CHAPTER 3
RO ISOLATION AND RESIN FRACTIONATION OF DOM AND THEIR IMPACT ON HAA AND THM FORMATION

INTRODUCTION

In this chapter, the effectiveness of RO isolation and resin fractionation of DOM and the impact of these methods on the reactivity of DOM toward THM and HAA formation will be discussed.

Characterization of Source Waters

The general water quality characteristics of each water source used in this study were examined by collecting a composite sample directly from the source during the RO isolation period. Selected physicochemical parameters measured in the samples are reported in Table 3.1. Myrtle Beach (MB) water has a high $SUVA_{254}$ and DOC values, whereas Greenville (GV) and Spartanburg (SP) waters are low DOC and low $SUVA_{254}$ waters. The watershed of the Myrtle Beach WTP contains many forests and wetland areas. Our experiences at this site have shown that the RO isolate of this water had intense brownish color (Kitis et al. 2001, Kitis et al. 2002, Kilduff et al. 2004). Pedogenic humic substances that generally contain more aromatic moieties and lignin, more intense color and larger molecular weight components are expected to dominate in this source. On the other hand, Greenville and Spartanburg waters are from reservoirs where the DOM is expected to have a more aquagenic signature, more aliphatic moieties and low molecular weight components. The RO isolates that we have collected from these sources had a yellowish color (Kitis et al. 2001, Kitis et al. 2002, Kilduff et al. 2004). All three sources are soft waters, which is typical for surface waters in South Carolina. The bromide detected in the Myrtle Beach water is probably due to seawater intrusion.

RO ISOLATION

In this section, the results of RO isolation experiments will be presented for each source water.

Greenville Water

Two separate DOM isolations (GV1 and GV2) were performed at the Greenville drinking water treatment plant. The results of water volume balance and DOC and dissolved nitrogen (DN) mass balance calculations for these runs are shown in Table 3.2 and in Figure 3.1. During each isolation, approximately 6.500 L of water was processed, which required one week of field work. This high processing volume resulted from the very low DOC of the source water (i.e., ~1.0 mg/L) and the minimum DOM mass needed (~5 grams) for the experimental matrix to be performed with each source water.

During the first isolation (GV1), a hydrogen cation exchanger was employed. The goal was to exchange cations in water with hydrogen in order to reduce the amount of dissolved salts in the DOM concentrate before the subsequent resin fractionation. The pH of the DOM concentrate

Table 3.1
Selected physicochemical characteristics of source waters*

Parameter	Unit	Greenville†	Myrtle Beach	Spartanburg
DOC	(mg/L)	0.8	10.7	1.9
UV_{254}	(cm^{-1})	0.014	0.522	0.043
$SUVA_{254}$	(L/mg-m)	1.9	4.9	2.3
Bromide	(μg/L)	<MRL	68	<MRL
pH	(–)	6.9	7.0	7.2
Alkalinity (as $CaCO_3$)	(mg/L)	8	32	11
Calcium	(mg/L)	1.10	8.20	2.48
Magnesium	(mg/L)	0.41	2.05	1.08
Iron	(mg/L)	0.05	0.69	0.03
Aluminum	(mg/L)	1.44	0.14	0.01
Dissolved nitrogen	(mg N/L)	0.11	1.45	0.49
Nitrate	(mg N/L)	0.05	0.23	0.03
Nitrite	(mg N/L)	<MRL	<MRL	0.01
Ammonia	(mg N/L)	<MRL	0.04	0.02
Dissolved organic nitrogen	(mg N/L)	0.06	1.18	0.43
Ionic strength	(as mM NaCl)	0.49	3.94	0.43

MRL: Minimum reporting level.
* All values are for samples filtered using a pre-washed 0.45-μm filter (either Osmonics PES or Gelman Supor).
† Results for the GV2 sample.

in the feed reservoir ranged between 4.0 and 4.5 during the course of the isolation. The results of water balance calculations was almost 100%, which was consistent with the field observation that there was no significant water spill or leaks from the system during the one week of operation. The use of hydrogen exchanger had two negative impacts: (1) significant DOM losses from the permeate, and (2) high amount of DOM fouling to the membrane. Only 73.1% of the DOM as organic carbon (i.e., RO concentrate and NaOH wash) was recovered. The high DOM loss was attributed to the decreasing size of DOM, and decreasing repulsive electrostatic interactions between the membrane surface and the DOM molecules with decreasing pH. Ghosh and Schnitzer (1980) presented a conceptual model to describe the configuration of humic substances at different ionic strengths, pH and concentrations (Figure 3.2). In addition, previous membrane research has confirmed configurational changes in humic substances with decreasing pH (Brown 1975; Kwak and Nelson 1977; Kilduff and Weber 1992).

A second isolation (GV2) was conducted at the same source about one week after the completion of the first one using a sodium cation exchanger. The pH of the DOM concentrate always remained neutral during this isolation. As for GV1, the water balance calculation of GV2 showed was almost 100%. In contrast to GV1, the DOC mass balance calculations showed that during GV2 run, significantly lower amount of DOM was lost in the permeate where DOC were usually below the MRL of the measurements and less amount of DOM fouled the membrane. As a result, about 88% of the DOM applied to the system was recovered.

Table 3.2
Water, DOC, and DN mass balances for the two Greenville water RO isolations

Water balance

	GV1		GV2	
	Volume (L)	%*	Volume (L)	%*
Water applied	**6,496**	**100**	**6,579**	**100**
Permeates	6,447	99.2	6,553	99.6
Concentrate	22	0.3	21	0.3
Water volume accounted for	**6,469**	**99.6**	**6,574**	**99.9**

DOC mass balance

	GV1				GV2			
	DOC (mg/L)	Volume (L)	Mass (mg)	%*	DOC (mg/L)	Volume (L)	Mass (mg)	%*
DOM applied	1.13	6,496.7	**6,978.6**	100	0.78	6,579	**5,132.8**	**100**
RO concentrate	150.2	22.4	**3,366.1**	**48.2**	167.3	20.7	**3,563.0**	**67.4**
Permeates	0.19†	6,447.3	**1,200.0**	**17.2**	0.07†	6,553.4	**451.6**	**8.8**
NaOH wash #1	19.85	50	992.7	14.2	17.41	50.6	881.0	17.2
NaOH wash #2	5.08	146.6	746.1	10.7	1.22	125.5	152.8	3.0
NaOH wash total		196.6	**1,738.8**	**24.9**		176.1	**1,033.8**	**20.2**
Water wash #1‡	1.43	146.6	209.1	3.0	0.84	149.9	126.6	2.5
Water wash #2	0.68	146.6	99.9	1.4	0.35	74.8	26.4	0.5
Water wash #3	0.35	146.6	51.2	0.7	0.12	146.6	18.10	0.4
Water wash #4	0.35	146.6	51.7	0.7	NP	NP	NP	NP
Water wash #5	0.93	146.6	136.7	2.0	NP	NP	NP	NP
Water wash total		733	**448.6**	**7.8**		371.3	**171.1**	**3.4**
DOM accounted for			**6,854**	**98.2**			**5,119.7**	**99.7**
DOM recovered§			**5,104.9**	**73.1**			**4,596.8**	**87.6**

DN mass balance

	GV1				GV2			
	DN (mg/L)	Volume (L)	Mass (mg)	%	DN (mg/L)	Volume (L)	Mass (mg)	%
DN applied	0.12	6,496.7	**755.0**	100	0.11	6,579	**750.0**	100
RO concentrate	6.72	22.4	**150.6**	**19.9**	15.21	20.7	**314.9**	**42.0**
Permeates	0.06†	6,447.3	**399.8**	**53.0**	0.03†	6,553.4	**225.1**	**30.0**
NaOH wash #1	0.98	50	49.0	6.5	1.80	50.6	91.0	12.1
NaOH wash #2	NA	146.6	NA	NA	0.25	125.5	31.0	4.1
NaOH wash total		196.6	**49.0**	**6.5**		176.1	**122.0**	**16.2**
Water wash #1‡	0.18	146.6	26.8	3.5	0.18	149.9	27.7	3.7
Water wash #2	0.13	146.6	19.6	2.6	0.13	74.8	10.1	1.3
Water wash #3	0.15	146.6	21.4	2.8	0.12	146.6	17.4	2.3
Water wash #4	0.06	146.6	8.6	1.1	NP	NP	NP	NP
Water wash #5	0.13	146.6	18.4	2.4	NP	NP	NP	NP
Water wash total		733	**94.8**	**12.5**		371.3	**55.2**	**7.3**
DN accounted for			**694.2**	**92.0**			**717.2**	**95.6**
DN recovered			**199.6**	**26.4**			**436.9**	**58.2**

NA: Not available; NP: not performed.
* With respect to DOM applied.
† Represent the average of about 53 individual permeate collected during the run.
‡ DDW was used.
§ The sum of RO concentrate and NaOH wash that was brought to the laboratory.

The DN mass balances indicated that 92 and 96% of DN were accounted for during each run. However, the DN concentrations in the source water were about the MRL of the measurements. The DN concentrations in the permeate were below the MRL. Therefore, although good mass balances were obtained, not too much emphasis should be placed on these results due to very low level of DN in the Greenville water. DN mass balances were not broken down to dissolved organic and inorganic nitrogen because many samples were around or below the detection limit of the measurements.

The results of Greenville water showed that using a sodium exchanger (thus maintaining pH around 7) would be important to minimize the losses while isolating DOM from different sources. Therefore, it was decided to use the sodium exchanger in other source water isolations and GV2 RO isolate in the subsequent chloramination experiments.

Myrtle Beach Water

The DOM isolation was performed at the Myrtle Beach drinking water treatment plant during the last week of June 2004. The results are presented in Table 3.3 and Figure 3.3. As for Greenville, a good water balance was obtained. The DOC mass balance calculations showed that 99% of DOM was recovered during the isolation. This high recovery was probably due to high molecular weight nature of DOM components, as indicated by the high $SUVA_{254}$ value of this water. It has been reported that molecular weight of DOM components generally increases with increasing SUVA values (Chin et al. 1994). The higher DOM recovery from Myrtle Beach than Greenville water is consistent with our previous experiences in soft waters that higher DOM recoveries were obtained in waters with high SUVA and high DOC than those having low SUVA and low DOC (Kitis et al. 2001). It was also noted that only 10% of applied DOM mass was recovered during NaOH wash for the Myrtle Beach water. This amount was lower than the recovery (20%) observed for the Greenville water. It appears that the DOM components in the low-$SUVA_{254}$ Greenville water more effectively fouled the RO membrane than those in the high-$SUVA_{254}$ Myrtle Beach water.

Spartanburg Water

The DOM isolation was performed at the Spartanburg drinking water treatment plant in February 2005. The water and mass balance results are presented in Table 3.3 and Figure 3.4. As for the Greenville and Myrtle Beach waters, the water loss during the isolation was negligible; as a result, a good water balance was obtained. The DOC mass balance showed that 87% of DOM was recovered during the RO isolation (Table 3.3). This recovery was similar to that of Greenville water (87.6 % GV2) but lower than the Myrtle Beach water (99.3%).

Comparison of RO Isolations for Three Source Waters

Table 3.4 shows the $SUVA_{254}$ values of the raw water composites and RO concentrates for three water samples. A difference of about 0.5 to 1.0 unit was observed between the raw water composite and the RO concentrate for the two low SUVA samples. A combination of two factors may have contributed to this difference: 1) due to low DOC concentrations, isolation of DOM from the low $SUVA_{254}$ waters required significantly longer period (e.g., about a week) or greater volume of water to be processed as compared to Myrtle Beach water (Tables 3.1 and 3.2). During

Table 3.3
Water, DOC, and DN mass balances for the Myrtle Beach and Spartanburg waters RO isolations

Water balance

	Myrtle Beach water		Spartanburg water	
	Volume (L)	%*	Volume (L)	%*
Water applied	**2,388**	**100**	**7,149**	**100**
Permeates	2,338	97.9	7,156	100.1
Concentrate	56	2.3	43	0.6
Water volume accounted for	**2,394**	**100.2**	**7,199**	**100.7**

DOC mass balance

	Myrtle Beach water				Spartanburg water			
	DOC (mg/L)	Volume (L)	Mass (mg)	%*	DOC (mg/L)	Volume (L)	Mass (mg)	%*
DOM applied	12.61	2,388	**30,113.3**	**100**	2.21	7,149	**15,796.1**	**100**
RO concentrate	476.1	56.4	**26,853.9**	**89.2**	284.02	43.0	**12,220.6**	**77.5**
Permeates	0.18†	2,338	**427.2**	**1.4**	0.26	7,156.6	**1,891.1**	**12.0**
NaOH wash #1	73.12	41.7	3,052.6	10.1	28.42	48.9	1,389.6	8.8
NaOH wash #2	NA	NA	NA	NA	1.79	38.7	69.5	0.4
NaOH wash total		41.7	**3,052.6**	**10.1**		87.6	**1,459.1**	**9.2**
Water wash #1‡	4.25	74.6	317.1	1.1	3.07	66.5	204.0	1.29
Water wash #2	0.62	74.6	46.1	0.2	0.94	127.2	119.5	0.76
Water wash #3	NP	NP	**NP**	**NP**	0.43	54.5	23.7	0.15
Water wash total		149.3	**363.2**	**1.3**		**248.2**	**347.2**	**2.2**
DOM accounted for			**30,696.9**	**101.9**			**15,918**	**100.8**
DOM recovered§			**29,906.5**	**99.3**			**13,679.7**	**86.7**

DN mass balance

	Myrtle Beach water				Spartanburg water			
	DOC (mg/L)	Volume (L)	Mass (mg)	%*	DOC (mg/L)	Volume (L)	Mass (mg)	%*
DN applied	1.45	2,388	**3,458.2**	**100**	0.44	7,149	**3,132.9**	**100**
RO concentrate	47.41	56.4	**2,674.1**	**77.3**	40.98	43.0	**1,763.6**	**56.3**
Permeates	0.20†	2,338.3	**464.98**	**13.4**	0.14	7,156.6	**1,019.2**	**32.5**
NaOH wash #1	7.48	41.7	312.2	9.0	5.02	48.9	245.4	7.8
NaOH wash #2	NA	NA	NA	NA	0.38	38.7	14.7	0.5
NaOH wash total		41.7	**312.2**	**9.0**		87.6	**260.1**	**8.3**
Water wash #1‡	0.47	74.6	35.3	1.0	0.88	66.5	58.3	1.9
Water wash #2	0.05	74.6	3.5	0.1	0.24	127.2	30.8	1.0
Water wash #3	NP	NP	**NP**	**NP**	0.15	54.5	8.2	0.2
Water wash total		149.2	**38.8**	**1.1**		248.2	**97.3**	**3.1**
DN accounted for			**3,490.1**	**100.9**			**3,140.2**	**100.2**
DN recovered			**2,986.3**	**86.3**			**2,023.7**	**64.4**

NA: Not available; NP: not performed.
* With respect to DOM applied.
† Represent the average of about 46 individual permeate collected during the run.
‡ DDW was used.
§ The sum of RO concentrate and NaOH wash that was brought to the laboratory.

Table 3.4
$SUVA_{254}$ values for raw composites and RO concentrates and percent DOM losses during RO isolations*

	GV2	SP	MB
		$SUVA_{254}$ (L/mg-m)	
Raw composite	1.9	2.3	4.9
RO concentrate	2.4	3.2	4.9
		% DOM losses from the RO concentrate	
Permeate	8.8	12.0	1.4
To membrane (NaOH wash)	20.1	9.2	10.1
DI wash	3.3	2.2	1.2
Total loss	32.2	23.4	12.7

* All three isolations were performed at about pH 7.

these long isolation periods, rain events have occurred at both sources. For the creation of a raw water composite, one sample was collected everyday around noon time. Therefore, the composite sample obtained during about one week of isolation was not a true time or flow proportional composite sample. It is possible that the variability in the source water $SUVA_{254}$, especially during the rain events, was not completely captured in the raw water composites, and 2) the DOM losses from the permeates ranged 9 to 12% for the two low-SUVA waters. It is likely that the DOM passing through the membrane is composed of mainly small molecular weight low UV absorbing components. As a result, the $SUVA_{254}$ values of RO concentrates can be higher than those of the raw water composite samples.

There were major differences among the three RO isolations in terms of the percent of DOM mass lost in permeates and the percent of DOM mass recovered from the membrane during NaOH wash (Table 3.4). For the two low $SUVA_{254}$ samples, the DOM lost in the permeates constituted 8.8 and 12% of the total DOM mass for Greenville and Spartanburg waters, respectively. The DOM mass recovered during the NaOH wash from the membranes was 20.1 and 9.2% for these waters, respectively. While a higher amount of DOM was lost in the permeates of Spartanburg water as compared to Greenville water, the membrane fouling was higher in Greenville than Spartanburg water. Understanding the cause of such interactions and its relationships to DOM and membrane material characteristics was beyond the scope and objective of this project. For example, Leenheer et al. (2000) found that DOM fouling of the membranes during the RO concentration of the low-$SUVA_{254}$ Colorado River water was due to bacterial mucopolysaccharides. Although it was not investigated in this study, similar type of DOM components may have been present in the Greenville water causing high degree of membrane fouling. For the high SUVA Myrtle Beach water, only 1.4% of DOM was lost in the permeates and only 10% of applied DOM mass was recovered during the NaOH wash. The low degree of loss to the permeate was probably due to high molecular weight nature of DOM components in this water, as indicated by its high $SUVA_{254}$ value. It is also evident that $SUVA_{254}$ values did not show a strong correlation with the extent of individual DOM-membrane interactions (i.e., losses to or through the membrane), however the total DOM losses during RO isolation showed an increasing trend with decreasing $SUVA_{254}$.

IMPACT OF RO ISOLATION ON DBP REACTIVITY

The general approach to evaluate the impact of RO isolation on the reactivity of DOM toward the THM and HAA formation for a source water was to chlorinate four samples side-by-side at the same UFC conditions. These experiments were performed with one low-$SUVA_{254}$ (i.e., Greenville) and one high $SUVA_{254}$ (Myrtle Beach water). The first sample (i.e., Raw-Filtered) was the filtered (using 0.45-µm pre-washed filter) composite raw water sample that was collected during the RO isolation directly from the source. The second sample (i.e., RO concentrate) was the RO isolate of the same source but diluted to the DOC level of the raw water. The third solution (i.e., NaOH wash) was prepared from the NaOH wash water (neutralized at pH~7) at the same DOC concentration of the raw water. The goal was to compare the DBP reactivity of membrane-fouling DOM components with the reactivity of RO concentrate (i.e., DOM in solution). Finally, the fourth solution (i.e., Reconstituted) was prepared at the same DOC of the raw water by mixing the RO concentrate and NaOH wash water (neutralized at pH~7). Since the RO concentrate and NaOH wash water did not always constitute 100% of the overall DOM recovered from the source (due to losses to the permeate and DDW washes), the percent contribution of RO concentrate and NaOH wash water were proportionally increased when preparing the reconstituted sample at the same DOC level of the raw water. Chlorine was used in the UFC tests since the DOC concentration of Greenville water studied in this project was very low; as a result chloramination would form trace amount of DBPs. However, one test was conducted using Myrtle Beach water (~10 mg DOC/L) and chloramines to check the preservation of chloramine reactivity. At the end of the UFC test, the formation and speciation of THMs and HAAs in all samples were measured and compared. In addition, DOC, Cl_2 consumption, pH at the end of the experiment, and $SUVA_{254}$ were determined for each solution used in these experiments. This approach is called the "reactivity check test" for RO isolation.

The results of the reactivity check tests are provided in Table 3.5 and Table 3.6. No significant difference was observed in the THM and HAA formation and speciation between the raw water, reconstituted raw water, NaOH wash water and RO concentrate using both chlorine and chloramines. This was consistent with our previous experience during the chlorination of waters in South Carolina (Kitis et al. 2001). In order to evaluate the reactivity of the DOM fraction that was not recovered during the RO isolation, a sensitivity analysis was performed by assuming that the missing DOM had from zero, ¼, ½, one, and two times the reactivity of the recovered DOM. The results are presented in Table 3.7. For chlorination of the Greenville water, the calculated true reconstituted THM reactivities ranged from 31 to 39 µg/mg, and HAA reactivities ranged from 33 to 41 µg/mg. The measured reconstituted water THM and HAA reactivities were 35 and 37 µg/mg, respectively. For chlorination of Myrtle Beach water, the calculated true reconstituted THM reactivities ranged from 39 to 41 µg/mg and HAA reactivities ranged from 53 to 57 µg/mg, as compared to the reconstituted water THM and HAA reactivities of 40 and 55 µg/mg, respectively. Because the missing DOM constitutes only 12 or 3% of the original DOM, it appears that even if they had significantly higher reactivities compared to the recovered DOM, it would not make a major difference in the reactivity of the reconstituted DOM considering the analytical variability of measurements. Therefore, for such small DOM losses, it is not possible to conclusively judge the reactivity of the missing DOM. The results of chloramination test results were also consistent with those of chlorination.

Table 3.5
Chlorine reactivity check experiments for Greenville and Myrtle Beach waters*

	Water	DOC (mg/L)	Cl_2 applied (mg/L)	Cl_2 residual (mg/L)	pH†	$SUVA_{254}$ (L/mg-m)	TCM (µg/mg)	BDCM (µg/mg)	THM (µg/mg)	DCAA (µg/mg)	TCAA (µg/mg)	BDCAA (µg/mg)	HAA (µg/mg)
Greenville	Raw—filtered	0.9	1.9	1.1	7.0	2.2	28	4	32	17	17	3	37
	Reconstituted	1.0	1.9	1.0	7.1	2.4	30	5	35	16	18	3	37
	RO concentrate	1.0	1.9	1.1	7.1	2.4	29	5	34	16	18	3	37
	NaOH wash	0.9	1.9	1.1	7.1	2.3	27	5	32	16	18	3	37
Myrtle Beach	Raw—filtered	10.5	15.0	0.8	7.2	5.2	37	2	39	17	39	2	59
	Reconstituted	9.4	12.8	0.8	7.2	5.1	38	2	40	17	36	2	55
	RO concentrate	9.5	12.8	0.8	7.2	5.1	40	2	42	17	36	2	55
	NaOH wash	10.1	15.0	1.1	7.2	5.4	38	2	40	17	40	2	60

* The concentration of other THM and HAA species were below the MRL. The results represent the average of two independent samples run for each water.
† Measured at the end of chlorination.

Table 3.6
Chloramine reactivity check experiments for Myrtle Beach water*

Water	DOC (mg/L)	Cl_2 applied (mg/L)	Cl_2 residual (mg/L)	pH†	$SUVA_{254}$ (L/mg-m)	TCM (µg/mg)	BDCM (µg/mg)	THM (µg/mg)	DCAA (µg/mg)	TCAA (µg/mg)	BDCAA (µg/mg)	HAA (µg/mg)
Raw—filtered	10.5	3.3	1.7	7.2	5.2	<MRL	<MRL	<MRL	1.7	<MRL	<MRL	1.7
Reconstituted	9.4	3.3	1.7	7.2	5.1	0.1	<MRL	0.1	1.9	<MRL	<MRL	1.9
RO concentrate	9.5	3.3	1.7	7.2	5.1	0.1	<MRL	0.1	1.9	<MRL	<MRL	1.9
NaOH wash	10.1	3.7	1.6	7.2	5.4	0.1	<MRL	0.1	2.0	<MRL	<MRL	2.0

* The concentration of other THM and HAA species were below the MRL. The results represent the average of two independent samples run for each water.
† Measured at the end of chloramination.

Table 3.7
Significance of missing DOM reactivity: A sensitivity analysis

			Greenville water—Chlorination			
	Percent			THM Reactivity (μg/mg)		
Raw water	100	32	32	32	32	32
Reconstituted	88	35	35	35	35	35
Hypothetical missing DOM	12	0	9	18	35	70
True reconstituted	100	31	32	33	35	39
	Percent			HAA Reactivity (μg/mg)		
Raw water	100	37	37	37	37	37
Reconstituted	88	37	37	37	37	37
Hypothetical missing DOM	12	0	9	19	37	74
True reconstituted	100	33	34	35	37	41
			Myrtle Beach water—Chlorination			
	Percent			THM Reactivity (μg/mg)		
Raw water	100	39	39	39	39	39
Reconstituted	97	40	40	40	40	40
Hypothetical missing DOM	3	0	10	20	40	80
True reconstituted	100	39	39	39	40	41
	Percent			HAA Reactivity (μg/mg)		
Raw water	100	59	59	59	59	59
Reconstituted	97	55	55	55	55	55
Hypothetical missing DOM	3	0	14	28	55	110
True reconstituted	100	53	54	54	55	57
			Myrtle Beach water—Chloramination			
	Percent			THM Reactivity (μg/mg)		
Raw water	100	<MRL	<MRL	<MRL	<MRL	<MRL
Reconstituted	97	0.1	0.1	0.1	0.1	0.1
Hypothetical missing DOM	3	0.0	0.0	0.1	0.1	0.2
True Reconstituted	100	0.1	0.1	0.1	0.1	0.1
	Percent			HAA Reactivity (μg/mg)		
Raw water	100	1.7	1.7	1.7	1.7	1.7
Reconstituted	97	1.9	1.9	1.9	1.9	1.9
Hypothetical missing DOM	3	0.0	0.5	1.0	1.9	3.8
True Reconstituted	100	1.8	1.9	1.9	1.9	2.0

Several conclusions can be drawn from these observations and analyses:

- The DOM components lost during the RO isolation through the permeate and DDW washes constituted about 12% and 3% of the overall DOC masses for Greenville and Myrtle Beach waters, respectively. Since the missing DOM comprises only a small fraction of the overall DOM, it was not possible to judge their reactivity compared to the recovered DOM with the analytical variability of measurements.
- The DOM captured with NaOH wash and the DOM in the RO concentrate had relatively similar $SUVA_{254}$ values, chlorine or chloramine demands and reactivities toward THM and HAA formation.
- Although bromide concentrations were not measured in each sample used for the reactivity check experiments, the absence of formation of brominated THM and HAA species in the raw (and other three) samples is consistent with the low-bromide concentrations (more importantly Br^-/DOC ratios) measured in the source waters.
- Similar chlorine consumptions, THM and HAA yields and speciation, $SUVA_{254}$ values of raw, reconstituted and RO concentrate waters showed that RO isolation maintained the reactivity of DOM in the source waters toward THM and HAA formation for the two waters tested in the study.

RESIN FRACTIONATION OF DOM

In this section, the results of DOM fractionation experiments will be presented. Fractionation of Greenville and Myrtle Beach water DOM was performed. It was decided to use the RO concentrate in the fractionation for two major reasons: (1) the RO concentrate and NaOH wash samples for each water did not show significantly different reactivities. Both the RO concentrate and NaOH wash had relatively similar $SUVA_{254}$ values. There was no apparent indication of preferential fractionation of DOM components between the RO concentrate and NaOH wash, and (2) it was experimentally more feasible to use only the RO concentrate, otherwise a large volume of NaOH wash should have been concentrated before fractionation.

During a fractionation experiment, a 20-L RO concentrate of 150 mg/L DOC concentration at pH 2 was prepared and applied to the XAD-8 resin. The permeate from the XAD-8 resin was collected, and its volume and DOC concentration was measured. Then the XAD-8 permeate was applied to the XAD-4 resin bed. The volume and concentration of DOC of the XAD-4 permeate (i.e., HPI fraction) were measured. These measured volumes and DOC concentrations were used to calculate the distribution of DOM in different fractions. At the end of the fractionation experiments, the HPO and TPH fractions were rotoevaporated, freeze-dried, and stored in a desiccator. The HPI fraction was kept in water at pH 4 and stored in a refrigerator at 4°C until use.

Greenville Water

The mass distribution of different DOM fractions in the Greenville water is presented with a pie chart format in Figure 3.5. The results showed that Greenville DOM consisted of 61% HPO, 17% TPH, and 22% HPI fractions. The percentage of HPO fraction was high for a low-$SUVA_{254}$ sample. During the resin fractionation (at pH 2), formation of some colloidal species was observed. It was suspected that some of these colloidal particles might be silica. The ICP/MS analysis of the raw water and RO concentrate showed that the silica (SiO_2) concentration in these

Table 3.8
$SUVA_{254}$ values* of the DOM fractions of Greenville and Myrtle Beach waters

Fraction	Greenville	Myrtle Beach
HPO	3.27	5.02
TPH	1.22	2.19
HPI	1.87	2.16

* L/mg-m.

samples were 39 and 8104 mg/L, respectively. During the RO isolation 6579 L of raw water was concentrated to 21 L, providing a concentration factor of 313 ($V_{RO\ concentrate}/V_{Applied}$). If all the silica in the raw water was concentrated, the concentration of silica in the RO concentrate would be about 12335 mg/L. The measured concentration of 8104 mg/L indicated that about 65% of silica in the raw water was concentrated in the RO isolate. This concentration was significantly higher than the solubility limits of amorphous silica ($SiO_2(s)$) in DDW. Calculations showed that the solubility of silica is 119 mg/L at pH 2 ($K_{sp} = 10^{-2.7}$) (Stumm and Morgan 1996). Although the presence of DOM and cations may impact the solubility of silica, some fraction of silica precipitated with the HPO fraction (at pH~2) in the XAD-8 column. These colloidal precipitates may have caused some hydrophilic DOM (e.g., TPH and HPI) components captured in the XAD-8 column yielding a higher percent of HPO fraction. It has been shown that adsorption of humic acids by silica increases significantly with decreasing pH (e.g., Fairhurst et al. 1995). These findings showed that selection of concentration factor during RO isolation with respect to the silica level in a source water is important in the subsequent fractions of DOM with XAD columns.

Myrtle Beach Water

The DOM fractionation experiment showed that Myrtle Beach water DOM consisted of 73% HPO, 17% TPH and 11% HPI fractions (Figure 3.6). This distribution is consistent with the high $SUVA_{254}$ value of this water. The concentration factor used for this water was 42 due to high raw water DOC concentration. The results of silica measurements of the RO concentrate was 160 mg/L, which was significantly lower than that of Greenville water.

DISTRIBUTION AND CHARACTERISTICS OF DOM FRACTIONS

Table 3.8 shows the $SUVA_{254}$ values of the Greenville and Myrtle Beach water DOM fractions. The hydrophobic fractions, as expected, had the highest $SUVA_{254}$ values. While the TPH fraction had a $SUVA_{254}$ lower than the HPI fractions in Greenville water, the $SUVA_{254}$ values of Myrtle Beach TPH and HPI fractions were similar. During the cleaning of the resins after fractionation, a brownish color was observed in the NaOH and DDW wash solutions, which disappeared with increasing wash volume. The color was more darker in the Myrtle Beach wash solutions than that of Greenville water. Therefore, there was visual evidence that some DOM components were not completely captured back from the resins during the fractionation process. The resins were washed several times until the TOC concentration of the wash solution was reduced below 1 mg/L. It was not feasible to quantify the amount of the DOM lost during the fractionation due to the extensive volume of water that needs to be collected and analyzed, as well as exhaustive amount of meticulous work that would have been required. For example, the presence of acetonitrile, the solvent used for elution from the resins, in some of these solutions would require its removal (e.g., extensive rotaevaporation, as performed for the obtained fractions) for proper DOC measurements. Using the $SUVA_{254}$ values of fractions and the distribution of HPO, TPH and HPI fraction in each water (Figure 3.5 and Figure 3.6), the $SUVA_{254}$ values of the whole waters were calculated and compared with the measured values of the RO concentrates. It was found that $SUVA_{254,calculated}$ of Greenville and Myrtle Beach water are 2.6 and 4.2 L/mg-m, while $SUVA_{254}$ of their RO concentrates were about 2.4 and 5.1 L/mg-m, respectively. Therefore, it appears that some DOM components, especially high UV absorbing fractions, of Myrtle Beach water were not completely captured during the fractionation process.

IMPACT OF DOM FRACTIONATION ON DBP REACTIVITY

To evaluate the impact of resin fractionation on the THM and HAA reactivity of Myrtle Beach and Greenville DOMs, a reconstituted water was prepared for each water by mixing the HPO, TPH and HPI fractions according to their mass distributions in the raw waters, as determined from the fractionation experiments (Figure 3.5 and Figure 3.6). The reconstituted sample was prepared at about the same DOC concentration as the raw water. In addition, the RO concentrate of the same water was diluted to the DOC of the raw water to prepare a fresh RO concentrate. Previous experiments showed that RO concentration did not result in any change in the THM and HAA formation reactivity (Table 3.5 and Table 3.6). Therefore, the new RO concentrate served as a control (i.e., representative of the raw water) for the experiment. Another objective of running the RO concentrate was to compare the results of this new RO concentrate with the one obtained about 2 months earlier to examine if the DOM reactivity of the RO concentrate changed during storage in the dark at 4°C in a walk-in refrigerator. Because the RO concentrate served as a stock to prepare solutions, it was important to monitor its reactivity during the storage period. The two samples, reconstituted and RO concentrate, were chlorinated and chloraminated side-by-side under the UFC conditions.

The results are presented in Table 3.9. The first comparison was made between the results of new (Table 3.9) and old (Table 3.5 and Table 3.6) RO concentrates. For Greenville water, both RO concentrates showed very similar THM and HAA formation and speciation, indicating that the reactivity did not change during the storage time. The $SUVA_{254}$ and chlorine consumption of the new RO concentrate was higher than the first RO concentrate by 0.5 unit and about 0.3 mg/L, respectively. However, these differences did not result in any significant impact on the THM and HAA formation and speciation. The reconstituted water, despite its similar $SUVA_{254}$ and chlorine consumption as the RO concentrate, yielded 12 and 23% lower THM and HAA formations, respectively, compared to the RO concentrate. This lower DBP yield was attributed to the loss of some DOM components during the resin fractionation. However, the results showed that more than 77% of THM and HAA formation in the source was captured by the collected fractions. Furthermore, similarity between the THM and HAA speciation of reconstituted and RO concentrate samples suggested that the fractionation technique relatively maintained the reactivity of DOM toward THM and HAA formation.

For the Myrtle Beach water, the $SUVA_{254}$ and chlorine consumption of the new RO concentrate was very similar to that of the old RO concentrate. Both RO concentrates showed very similar HAA formation and speciation, but THM formation, especially TCM, of the new RO concentrate was 6 μg/L lower (i.e., 14%) as compared to the old RO concentrate. Despite the

Table 3.9
Impact of DOM fractionation on the DOM reactivity toward THM and HAA formation*

Greenville

Water	DOC (mg/L)	Cl_2 applied (mg/L)	Cl_2 residual (mg/L)	pH†	$SUVA_{254}$ (L/mg-m)	TCM (µg/mg)	BDCM (µg/mg)	THM (µg/mg)	DCAA (µg/mg)	TCAA (µg/mg)	BDCAA (µg/mg)	HAA_9 (µg/mg)
Reconstituted	1.1	2.4	1.3	7.4	2.8	24	4	28	11	13	2	27
RO concentrate	1.1	2.4	1.3	7.4	2.9	27	5	32	14	18	2	35

Myrtle Beach

Water	DOC (mg/L)	Cl_2 applied (mg/L)	Cl_2 residual (mg/L)	pH†	$SUVA_{254}$ (L/mg-m)	TCM (µg/mg)	BDCM (µg/mg)	THM (µg/mg)	DCAA (µg/mg)	TCAA (µg/mg)	BDCAA (µg/mg)	HAA_9 (µg/mg)
Reconstituted	9.6	13.5	1.25	7.1	3.9	33	2	35	14	33	3	50
RO concentrate	9.6	13.5	0.88	7.1	5.1	34	2	36	17	39	2	58

Water	DOC (mg/L)	NH_2Cl_2 applied (mg/L)	NH_2Cl_2 residual (mg/L)	pH‡	$SUVA_{254}$ (L/mg-m)	TCM (µg/mg)	BDCM (µg/mg)	THM (µg/mg)	DCAA (µg/mg)	TCAA (µg/mg)	BDCAA (µg/mg)	HAA_9 (µg/mg)
Reconstituted	9.6	3.5	1.85	7.1	3.9	0.3	0.1	0.4	1.9	ND	ND	1.9
RO concentrate	9.6	3.5	1.65	7.1	5.1	0.3	0.1	0.4	2.4	ND	ND	2.4

* The concentration of other THM and HAA species were below the MRL.
† Measured at the end of chlorination.
‡ Measured at the end of chloramination.

small difference in the THM results, $SUVA_{254}$, chlorine consumption and HAA results overall suggested that there was no considerable change in the DBP reactivity of Myrtle Beach RO concentrate during storage. The reconstituted Myrtle Beach water had a significantly lower $SUVA_{254}$ value as compared to the RO concentrate (Table 3.9). As discussed in the previous section, some of the high UV absorbing DOM components of Myrtle Beach water were not completely captured during the fractionation process. Despite the significant difference between the $SUVA_{254}$ values of the reconstituted and RO concentrate DOM, their THM formation were similar under both chlorination and chloramination conditions, while a lower degree of HAA formation (14 and 20%) was observed in the reconstituted water than in the RO concentrate. This indicated that the DOM components lost during fractionation (i.e., high UV absorbing components) were mostly some HAA precursors. This is consistent with previous research that HAA precursors are more hydrophobic or exhibit higher UV_{254} absorption than THM precursors (Najm et al. 1994, Liang and Singer 2003).

During chlorination, the THM/HAA ratios were 35/50 = 0.70 and 36/58 = 0.62 for the RO concentrate and the reconstituted samples, respectively, for the high $SUVA_{254}$ Myrtle Beach DOM, and 28/27 = 1.03 and 32/35 = 0.91 for the low $SUVA_{254}$ Greenville DOM. This difference is probably not significant considering the analytical variability of measurements. Reckhow et al. (1990) found that the relative formation of THM and HAA, in particular between that of chloroform and TCAA, was an indicative of the nature of the precursors for a given DOM. For the reconstituted sample, the THM/TXAA ratio was 35/36 = 0.97, whereas for the RO concentrate the THM/TXAA ratio was 36/41 = 0.87 for the Myrtle Beach water, and 28/15 = 1.87 and 32/20 = 1.6 for Greenville water. These ratios were also comparable. These analyses further indicated that the fractions obtained for these waters captured most of the DBP reactivity of DOM in the source waters.

Overall, it is not surprising to have some degree of DOM loss occurring during the exhaustive resin fraction procedure. Despite these losses, the results obtained with one low-SUVA and one-high SUVA water investigated in this project showed that a great majority of (>77%) of THM and HAA formation in the source was captured by the collected fractions.

SUMMARY OF IMPORTANT FINDINGS

Important findings from this phase of this study are as follows:

- RO isolation is an effective method for concentrating and isolating DOM from natural waters. Our experience in this study and in our previous work have shown that higher DOM recoveries are obtained in waters having high SUVA and high DOC values than those having low SUVA and low DOC. This is attributed to generally increasing molecular weight of DOM components with increasing SUVA values. In this study, it was possible to recover 87, 88 and 99% of organic carbon from the source waters of Spartanburg, Greenville and Myrtle Beach water treatment plants, respectively, during RO isolation.
- A hydrogen or sodium exchanger prior to an RO system can be used to remove multivalent cations from water. The hydrogen exchanger reduces the cation content (except hydrogen) in the RO concentrate. However, the use of hydrogen exchanger significantly increases DOM losses (to both permeate and membrane) during RO concentration. This is due to the decreasing size of DOM and decreasing repulsive electrostatic interactions between the membrane surface and the DOM molecules with

decreasing pH when a hydrogen exchanger is used. In this study, for Greenville water, 74 and 88% of DOM recoveries were obtained using a hydrogen and sodium exchangers, respectively. The disadvantage of using a sodium exchanger is the presence of high sodium concentration in the RO concentrate, which can be problematic depending on the objective of subsequent use of the RO concentrate. Because the presence of high background sodium concentration was not an important factor in this study, a sodium exchanger was used in RO isolation.

- The DOM losses during RO isolation occur either through the permeate or to the membrane (i.e., fouling of the membrane). In this study, higher degrees of DOM losses through permeate were observed for low $SUVA_{254}$ than high $SUVA_{254}$ waters. This was attributed to the high molecular weight nature of DOM components in the latter than the former, as indicated by the $SUVA_{254}$ values. However, no correlation has been observed between the $SUVA_{254}$ values of the waters and the degrees of losses to the membrane (characterized as NaOH wash). On the other hand, the total DOM losses (i.e., permeate and membrane) showed an increasing trend with decreasing $SUVA_{254}$ values of the raw water.
- "Reactivity check" experiments showed that there was no impact of RO isolation on the DOM reactivity toward THM and HAA formation and speciation. In addition, DBP formation of RO isolates did not change during the 2–4 months of storage period in a 4°C refrigerator in the dark.
- Because both organic and inorganic matter are concentrated during RO isolation, the concentration of inorganic components, especially silica, may be problematic during the subsequent resin fractionation of DOM. Therefore, the concentration factor is an important consideration for RO isolation. In this study a high concentration factor (313) was used for the low DOC Greenville water to collect enough DOM mass from the source water. However, some degree of silica precipitation was observed as a result of adjusting the pH of RO concentrate to 2 prior to XAD-8 column fractionation. As a result, some of the TPH and HPI components in the Greenville water might have been captured in the HPO fraction because the adsorption of DOM by silica may become significant at low pH. Silica was not a problem for the Myrtle Beach sample due to the low concentration factor needed for its concentration.
- Resin fractionation results showed that Greenville and Myrtle Beach waters are composed of 61% HPO, 17% TPH, 22% HPI, and 72% HPO, 17% TPH, 11% HPI, respectively. The HPO fractions had the highest $SUVA_{254}$ values for both samples ($SUVA_{254,GV,HPO}$ = 3.3 and $SUVA_{254,MB,HPO}$ = 5.0 L/mg-m). While the TPH fraction had a $SUVA_{254}$ lower than the HPI fractions in Greenville water ($SUVA_{254,GV,TPH}$ = 1.2 and $SUVA_{254,GV,HPI}$ = 1.9 L/mg-m), the $SUVA_{254}$ values of Myrtle Beach TPH and HPI fractions were the same ($SUVA_{254,MB,TPH}$ = 2.2 and $SUVA_{254,MB,HPI}$ = 2.2 L/mg-m). Visual observations (e.g., a brown color in NaOH wash solutions) during resin cleaning after fractionation indicated that some DOM components were not completely recovered from the resins. Calculation of the $SUVA_{254}$ of the whole water from the individual $SUVA_{254}$ values of the fractions and comparison with the measured value of the whole water indicated that, especially for Myrtle Beach water, high UV absorbing components were not completely recovered from resins during fractionation.
- In order to quantify the impact of DOM losses during fractionation, reactivity check experiments were conducted side by side at the same DOC concentration for the

original RO concentrate and the reconstituted RO concentrate prepared using the individual DOM fractions. The results using chlorine showed that the Greenville reconstituted water from the individual fractions yielded 12 and 23% lower THM and HAA formations, respectively, compared to those of RO concentrate. For the Myrtle Beach water, the THM formation of the RO concentrate and reconstituted water was similar under both chlorination and chloramination conditions, while a lower degree of HAA formation (14 and 20%, respectively) was observed in the reconstituted water than in the RO concentrate. These results indicated that the DOM components lost during fractionation were mostly HAA precursors. Despite some of these losses, the results also showed that a great majority (i.e., >77%) of DOM reactivity toward THM and HAA formations in the source waters was captured by the collected fractions.

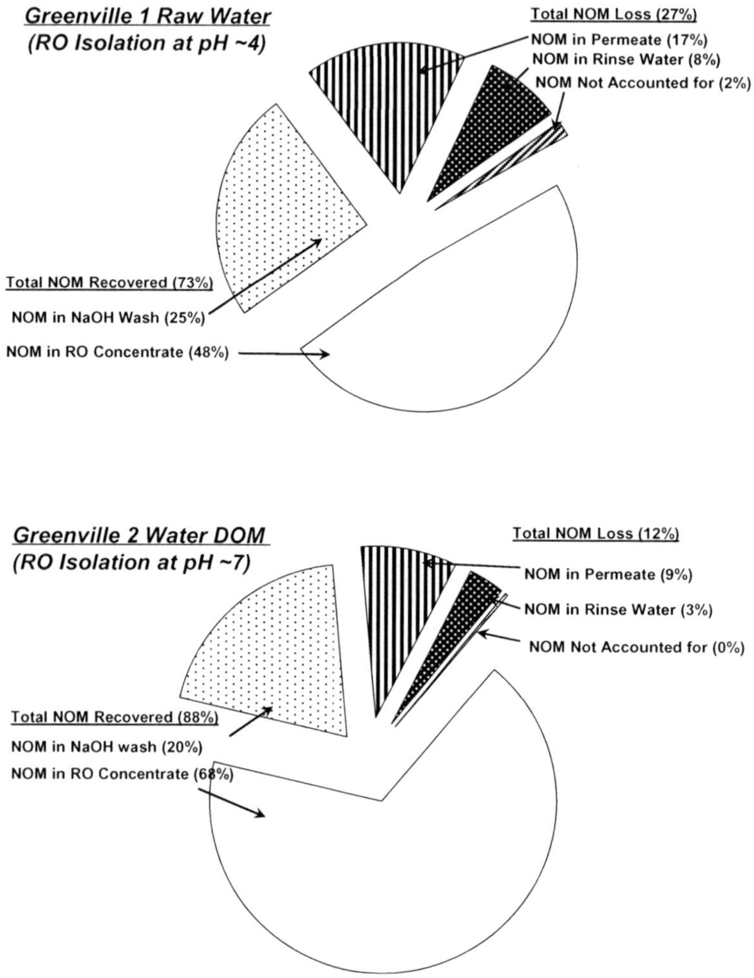

Figure 3.1 Distribution of DOM (as organic carbon mass) in Greenville water during two RO isolations

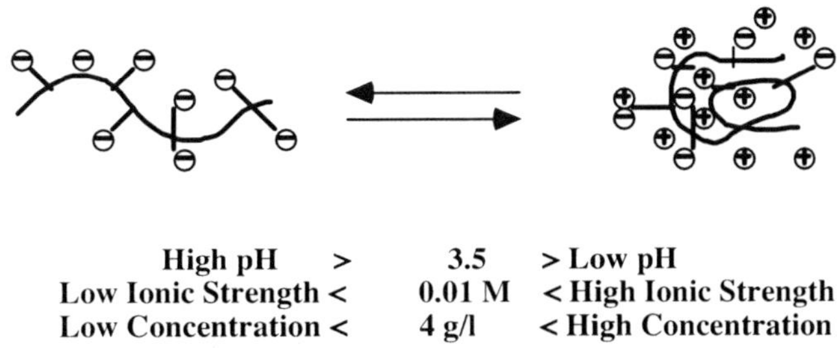

Figure 3.2 Configurational changes in DOM structure as a function of pH, ionic strength, and concentration

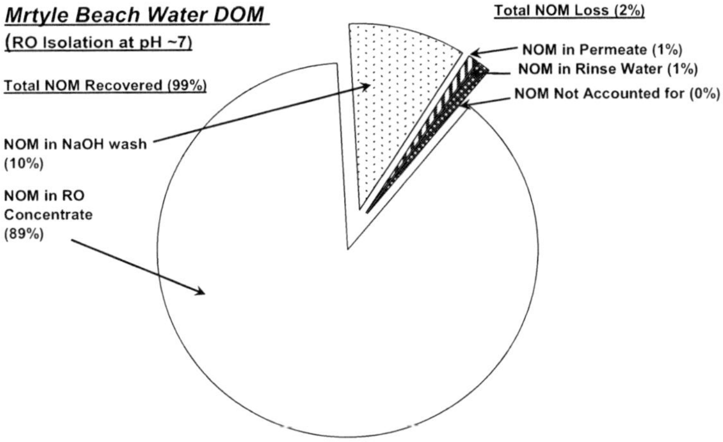

Figure 3.3 Distribution of DOM (as organic carbon mass) in Myrtle Beach water during RO isolation

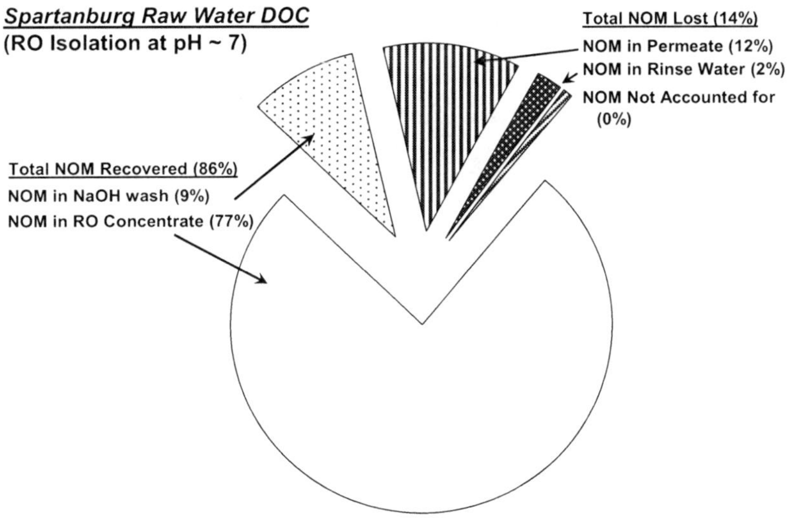

Figure 3.4 Distribution of DOM (as organic carbon mass) in Spartanburg water during RO isolation

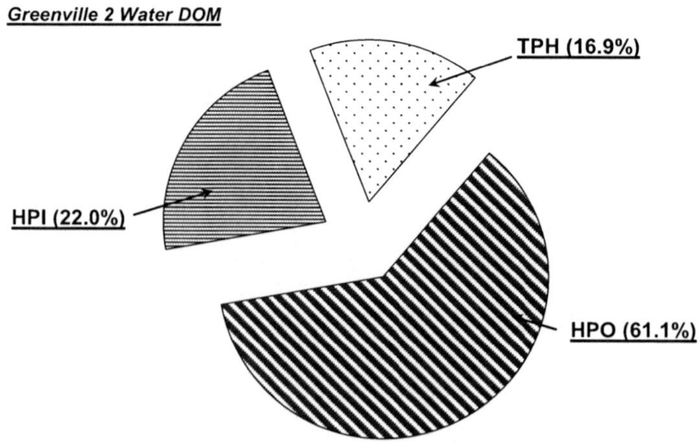

Figure 3.5 Distribution of DOM fractions in Greenville water

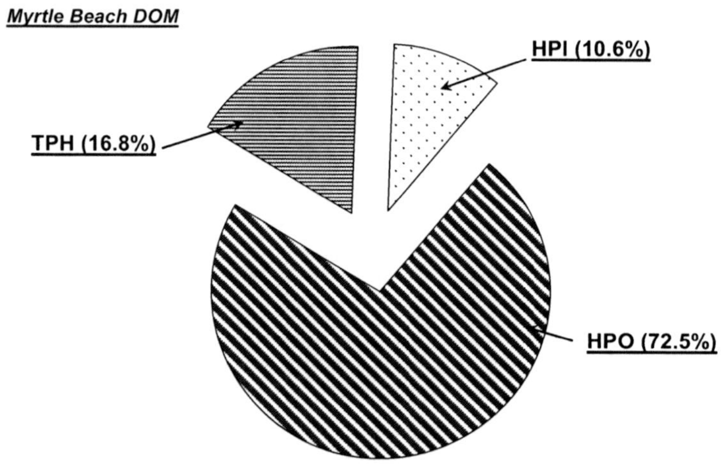

Figure 3.6 Distribution of DOM fractions in Myrtle Beach water

CHAPTER 4
HAA FORMATION PATHWAYS AND KINETICS DURING CHLORAMINATION AND THE PRACTICAL IMPLICATIONS

INTRODUCTION AND RESEARCH APPROACH

The two main objectives in this phase of this study were (1) to conduct a systematic investigation to gain insight into the formation and speciation of HAA, especially DXAA, and (2) to elucidate the cause(s) of inconsistent patterns reported in the literature for the HAA formation kinetics during chloramination. The results obtained for these two objectives will be presented in this chapter.

Three oxidants, NH_2Cl and its two decomposition products, HOCl and $NHCl_2$, can be present in water during chloramination, potentially reacting with DOM to form HAA. As discussed in detail in Chapter 1, until this project and the recent study by Duirk (2003), HAA formation during chloramination was commonly attributed to the formation of HOCl during NH_2Cl decomposition. The significance of a direct reaction between NH_2Cl and DOM on HAA formation has not been investigated. In this project, it was hypothesized that DXAA formation during chloramination results from a combination of both (i) direct reaction(s) of NH_2Cl with DOM, and (ii) reaction of HOCl, produced as a result of NH_2Cl decomposition, with DOM. Therefore, experiments were designed and performed to examine the relative significance of the direct reaction of NH_2Cl with DOM in the HAA formation compared to all the other complex reactions that can occur during chloramination. These experiments were conducted in the absence of bromide to minimize its impact (i.e., formation of bromamines and bromochloramines) on the already complicated chloramine chemistry. Furthermore, to date no study has examined the formation of HAA from $NHCl_2$, which was also a topic of investigation in this project. Therefore, experiments were conducted using preformed $NHCl_2$ to examine its role in HAA formation.

It is evident from the reactions listed in Table 1.1 in Chapter 1 that by increasing background ammonia concentration, autodecomposition of NH_2Cl, and thus HOCl production (reaction 1-2) and formation of $NHCl_2$ (reactions 1-3 and 1-5), can be significantly suppressed. When HOCl and $NHCl_2$ formation are minimized, the occurrence of reactions involving these species (reactions 1-4, 1-6 to 1-10) are likely to be minimized. Therefore, by performing experiments in the absence and presence of high background ammonia, it is possible to assess the relative significance of the direct reaction of NH_2Cl with DOM on HAA formation during chloramination. Another approach to evaluate the importance of the direct reaction is to conduct the experiments at a high pH where NH_2Cl decomposition significantly decreases. However, this approach does not provide information about the impact of pH on the HAA yields from the direct reaction of NH_2Cl with DOM since the experiments can only be conducted at high pH. Therefore, in this study, formation of HAA during chloramination was examined by performing kinetics experiments in the presence and absence of background ammonia at pH values of 6 and 9.

DETERMINATION OF BACKGROUND AMMONIA LEVEL

In order to evaluate the significance of the NH_2Cl pathway (i.e., the reaction of NH_2Cl with DOM forming HAA), a proper level of background ammonia should be selected to suppress NH_2Cl decomposition (reactions 1 and 3 in Figure 1.1) during the experiments. Preliminary experiments

Table 4.1
Effect of background ammonia on NH_2Cl decay and HAA formation at pH 6*

	Myrtle Beach water			Greenville water		
NH_3-N† (mg/L)	DCAA (μg/L)	NH_2Cl (mg/L)	$NHCl_2$ (mg/L)	DCAA (μg/L)	NH_2Cl (mg/L)	$NHCl_2$ (mg/L)
0	28	1.8	0.6	21	1.7	0.4
10	24	2.4	<MRL‡	18	2.1	<MRL
20	24	2.4	<MRL	18	2.2	<MRL
30	24	2.4	<MRL	18	2.2	<MRL
40	24	2.4	<MRL	16	2.2	<MRL
50	24	2.3	<MRL	16	2.3	<MRL
100	23	2.3	<MRL	17	2.3	<MRL

* Br/DOC = 4 and 12 μg/mg for Myrtle Beach and Greenville, respectively. Initial NH_2Cl = 3.8 mg/L for Myrtle Beach DOM and 3.5 mg/L for Greenville DOM.

† Additional amount of NH_3-N added in each bottle at the beginning of the experiment.

‡ MRL: Minimum reporting level. Free chlorine after 24-h reaction period was always below the MRL.

were conducted using MB and GV DOM and preformed NH_2Cl at pH 6 at several background ammonia concentrations. Since NH_2Cl decomposition decreases with increasing pH, the required background ammonia level determined for pH 6 can also be used for high pH experiments.

Table 4.1 shows the concentrations of DCAA, residual NH_2Cl, and $NHCl_2$ after 24-hr contact time in both waters. At background NH_3-N concentrations above 10 mg/L (i.e., 0.7 mM), the residual NH_2Cl concentrations and DCAA formation remained almost constant in both waters, which was 2.4 mg/L and 2.1 mg/L of residual NH_2Cl, and 24 μg/L and 18 μg/L of DCAA for MB and GV DOM, respectively. For both DOMs, no detectable $NHCl_2$ formation was observed at 10 mg/L and higher background NH_3-N concentrations. Equilibrium calculations showed that the amount of HOCl that can be produced from NH_2Cl decomposition decreased by two orders of magnitude with increasing background NH_3-N from 0 to 100 mg/L. Therefore, if NH_2Cl decomposition and HOCl and/or $NHCl_2$ formation were significant, it would be expected to observe their impacts on the residual NH_2Cl concentration and the amount of DXAA formed as a function of background ammonia levels. Constant concentrations of NH_2Cl and DXAA and the absence of $NHCl_2$ formation above 10 mg/L background NH_3-N indicated that decomposition of NH_2Cl which may lead to some HAA formation (through production of HOCl and/or formation of $NHCl_2$) was significantly suppressed; therefore, the measured DXAA concentrations in the presence of background ammonia were attributed to the direct reaction of NH_2Cl with DOM. The amounts of DXAA formed in the presence of background ammonia (20 mg/L results were used for the comparison) were about 70 to 80% of the levels observed in the absence of ammonia. These initial results indicated that direct reaction of NH_2Cl with DOM was the dominant pathway in the HAA formation at pH 6. Furthermore, these results indicated that 20 mg/L background ammonia was sufficient to suppress NH_2Cl decomposition. Thus, this concentration was selected for further experiments.

EXPLORING HAA FORMATION PATHWAYS DURING CHLORAMINATION

To examine HAA formation pathways during chloramination, parallel kinetics experiments were performed using each DOM solution under two different conditions at pH 6 and 9. For condition 1, preformed NH_2Cl solution was added to the DOM solution allowing all reaction pathways to occur. For condition 2, the same experiment for condition 1 was conducted by adding background NH_3-N of 20 mg/L in the DOM solution. The results of the experiments for condition 2 (where NH_2Cl hydrolysis and $NHCl_2$ formation were significantly suppressed) were compared with those of conditions 1 (where NH_2Cl decomposition can occur) to assess the relative importance of the direct reaction(s) of NH_2Cl with DOM on HAA formation. For both conditions, the same stock solution, DOC concentration of 7.5 mg/L was used in the experiments. During the experiments, residual NH_2Cl, formation of $NHCl_2$ and pH were monitored. Results are presented from Figures 4.1 to 4.4, and the data collected are provided in a tabular format in Appendixes A and B. In addition, NH_2Cl autodecomposition in the absence of DOM was also monitored using a synthetic background solution prepared with DDW and carbonate buffer.

pH 6 Results

The results of NH_2Cl decay in the presence and absence of DOM at two different background ammonia levels are shown in Figure 4.1. In the absence of DOM, NH_2Cl decomposition was almost completely suppressed by addition of ammonia. The two NH_2Cl decay profiles (i.e., dashed lines in Figure 4.1) started to diverge almost from the beginning of the experiment and the difference between them increased with time, indicating that the ammonia in water controlled the NH_2Cl decay during the course of the experiment. After 72-hr reaction period, residual NH_2Cl was 87% of the initial concentration with 20 mg/L background ammonia, while it was only 48% with no ammonia addition.

In the presence of DOM, the NH_2Cl decay patterns were different. DOM significantly accelerated NH_2Cl decay. NH_2Cl decay profiles started to diverge after about one hour and a higher and faster degree of NH_2Cl decay was observed in the absence of background ammonia in MB water (the high SUVA water) than in GV water (the low SUVA water) (Figure 4.1). This was consistent with the previous reports that NH_2Cl decomposition increases with increase in the $SUVA_{254}$ of the DOM (Valentine et al. 1998). Noticeably, at each ammonia level (i.e., 0 and 20 mg/L), the NH_2Cl decay profiles after the first 8-hr period in the presence of DOM were parallel to those observed in the absence of DOM. No $NHCl_2$ was detected with the addition of 20 mg/L background ammonia in the presence and absence of DOM, providing additional evidence that NH_2Cl decomposition was significantly suppressed with background ammonia addition. On the other hand, under the conditions without background ammonia addition, $NHCl_2$ concentrations of 0.6 mg/L (for MB DOM), 0.4 mg/L (for GV DOM) and 0.8 mg/L of (without DOM) were measured after 72-hr reaction time.

HAA formation under two different ammonia conditions is shown in Figure 4.2. DCAA was the primary HAA specie formed during these experiments. Consistent with a higher degree of NH_2Cl decay, a higher degree of HAA formation was observed in the absence than in the presence of background ammonia. The formation of HAA was gradual and continuous, with the difference in HAA formation increasing with time. The difference became appreciable after 4-hr reaction time due to the slow nature of HOCl production from NH_2Cl decay. Over the course of the experiment, the ratio of HAA concentrations under two different ammonia conditions was around 0.75 to 0.85

for the two DOMs. This indicated that even if NH_2Cl decomposition was significantly suppressed, a significant amount of HAA still formed from the direct reaction of NH_2Cl with DOM. Furthermore, such degree of HAA formation was observed in spite of the fact that the direct reaction may have also been suppressed to some extent in the presence of background ammonia since ammonia has been reported one of the major product of DOM oxidation of NH_2Cl (Vikesland et al. 1998). During these experiments, the formation of THM formation was also monitored. Only trace concentrations of chloroform, 2–3 μg/L, were observed in the MB water.

To assess the importance of $NHCl_2$ for HAA formation, preformed $NHCl_2$, ranging from 0.1 to 2 mg/L was reacted with each DOM at pH 6, a pH favoring $NHCl_2$ formation and stability for 24 hours. The results showed no HAA formation, indicating that $NHCl_2$ does not contribute to HAA formation under typical drinking water treatment conditions. Overall, these results show that direct reactions between NH_2Cl and DOM play the major role, constituting about 80% of HAA formation, while HOCl contributes at most about 20% of overall HAA formation.

pH 9 Results

The results of NH_2Cl decomposition in the presence and absence of DOM at two different background ammonia levels are shown in Figure 4.3. In the absence of DOM, NH_2Cl was very stable and its decay was negligible independent of the background ammonia level. The presence of DOM resulted in a higher degree of NH_2Cl decay than in the absence of DOM, while there was only a slight difference (~0.2 mg/L) between the NH_2Cl residuals of two different ammonia experiments for both DOMs. The NH_2Cl decay profiles remained parallel to each other without showing a continuous separation with time as was observed at pH 6.

Since NH_2Cl decay was minimal at pH 9, the formation of HAA was caused only by direct reaction between NH_2Cl and DOM. DCAA was the only specie formed. There was no difference in HAA formation (within 1 μg/L) between the two ammonia level experiments (Figure 4.4) for MB water. The difference in DCAA formation for GV water after 72 hours was less than 2 μg/L, which was about the variability in HAA measurements. All these observations suggested that the effect of ammonia on HAA formation was not important at pH 9. The DCAA concentration at pH 9 was less than 20% of the DCAA formed by direct reaction at pH 6, indicating that DXAA formation from direct reaction depended on pH. This is different from the trend observed during chlorination where DXAA formation is independent of pH.

CAUSE OF DIFFERENT HAA FORMATION KINETIC PATTERNS IN THE LITERATURE

One noticeable observation obtained during the pathways studies was the significantly different HAA formation kinetics patterns observed in this study as compared to others reported in the literature (Figure 4.5). As indicated in Chapter 1, Speitel and co-workers reported very fast HAA formation kinetics for surface waters (Pope et al. 2002, Pope et al., 2003, Speitel et al. 2004). For example, they reported 73% of the 72-hr DXAA concentration formed after 0.5 hour and 82% formed after 1 hour in Metedenock River source water at pH 8. Hua and Reckhow (2005a) reported similar fast kinetics of HAA formation for a surface water (85% of 72-hr yield formed in 30 min) at pH 7. Such very fast HAA formation kinetics reported during chloramination were unusual because no such fast formation kinetics have ever been reported even for chlorine which is (i) a much stronger oxidant than chloramines, and (ii) one of the decomposition products of NH_2Cl to which HAA formation during chloramination was attributed. Furthermore, despite the presence of DOM, the hydrolysis of NH_2Cl and its disproportionation reaction are slow especially at pH above 7.5. Singer et al. (1999) also reported relatively fast kinetics in chloramination (Figure 4.5). Under relatively similar conditions, in this study, only about 10% of 72-hr yield was observed in the first hour. Based on close examination of the experimental conditions used in these studies and using the understanding gained from the pathways results obtained in this project, it was hypothesized that the use of ammonium chloride (NH_4Cl) as a quenching agent required by USEPA Method 552.3 for the DBP samples is the main reason for the different HAA formation patterns in chloramination kinetic studies.

USEPA Method 552.3, approved for compliance monitoring under the D/DBP Rule, requires the use of NH_4Cl as a chlorine quenching agent for DBP samples (USEPA 2003). NH_4Cl is not a quenching agent for NH_2Cl, the typical form of chloramines used for water treatment. NH_4Cl significantly minimizes NH_2Cl decomposition without eliminating it from sample vials. The residual NH_2Cl can continue to react with the DOM and bromide until the sample is extracted. Therefore, if the sample is not extracted immediately, the use of NH_4Cl may create an artifact on the HAA determination in chloraminated samples and especially in kinetics experiments where all the samples for different reaction times are typically collected and processed together at the end of the experiments. Although NH_4Cl does not quench NH_2Cl, in this report "quenching" term will still be used for NH_4Cl.

The use of a proper quenching agent for chloraminated samples is important for several reasons: (i) obtaining correct results for HAA measurements, (ii) obtaining correct kinetics information about formation of HAA, and (iii) obtaining proper interpretation and modeling of HAA formation and speciation. To assess the significance of quenching agent selection on chloramination kinetics, two experiments were performed side by side at two pHs (6 and 9) using Myrtle Beach DOM at the ambient bromide level (Br/DOC = 4 μg/mg) and DOC at 7.5 mg/L. Chloraminated samples were quenched using either a stoichiometric amount of Na_2SO_3 or 100 mg/L NH_4Cl at the end of each contact time. The samples were immediately stored in a refrigerator (4–6°C) until extraction, which was performed for all samples together right after the last sample of the kinetic experiment (48 hr) was quenched. The results are presented in Figure 4.6 and provided in a tabular format in Appendix C.

The samples quenched with NH_4Cl showed significantly faster and higher HAA formation kinetics than the samples quenched with Na_2SO_3 at both pH values. However, the quenching effect was less pronounced at high pH. The difference between the two kinetic patterns (i.e., NH_4Cl vs. Na_2SO_3) was greater in the early reaction time. For the case of NH_4Cl, the bottles at the early reaction times had small amounts of HAA formed. However, since the NH_2Cl residual was not quenched as a result of the NH_4Cl addition, the residual had two more days (i.e., the storage time until the extraction) to react with DOM. On the other hand, for the case of Na_2SO_3, since the NH_2Cl residual is completely destroyed at the time of the quenching agent addition, a significantly lower amount of HAA formation was observed than those quenched with NH_4Cl, especially at the early reaction times. Pepich et al. (2004) studied NH_4Cl as a quenching agent for chloraminated samples with a DOC of about 4 mg/L at pH 8.1. The time span of the kinetics experiment was 28 days. All the samples were quenched by NH_4Cl. The control samples were extracted right away. The HAA formation at such conditions was very low, only 3.8 μg/L. After two days, it increased to 6 μg/L and stayed constant for the rest of the storage time. The quenching effect was relatively small probably due to the high pH conditions. Moreover, since the control samples were also quenched with NH_4Cl, the time that elapsed until the "immediate" extraction may have resulted in a small amount of positive artifacts.

To further examine the effect of the quenching agent, additional samples were run in parallel with the same DOM for the early stage (0.5 and 1 hour reaction time) of chloramination using NH_4Cl or Na_2SO_3. Samples were extracted either on the same day of the experiment (extracted within 5 hours while stored in the refrigerator until extraction) or two days later (after being kept in the refrigerator at 4–6°C). For the samples quenched with Na_2SO_3 there was no impact of two-day storage time on the measured HAA concentrations (Figure 4.7). For the samples quenched with NH_4Cl, the HAA concentration at each reaction time was higher than that measured using Na_2SO_3. Furthermore, the HAA concentration of the same sample increased during the two days of storage. It is evident that complete quenching of chloraminated samples is critical to obtain correct HAA results, especially for samples collected at short contact times.

Previously it was reported that a very high of excess Na_2SO_3 may cause the destruction of some brominated acetic acids under storage condition (Krasner et al. 1989). In this study, the dose of Na_2SO_3 used for samples was about 0.5 ppm of Na_2SO_3 above the stoichiometric amount required for quenching. Experiments were also performed to examine if a 0.5 to 1 ppm excess of Na_2SO_3 residual will have any impact on the stability of HAA in the sample vials during the storage period. For these experiments, 5 and 25 ppb HAA standards were prepared in DDW and spiked with Na_2SO_3 to give 0.5 and 1.0 ppm residual. One set of standards was also prepared with no Na_2SO_3 spike. The standards were extracted 3 days later after being stored in the refrigerator (4–6°C). The same experiments were performed both at pH 6 and 9. The results are presented in Table 4.2. It is evident that there was no impact of 0.5 to 1.0 ppm residual on the stability of HAA during the three days of storage period. These results confirmed that using 0.5 to 1.0 ppm Na_2SO_3 residual above the stoichiometric amount required for quenching will provide reliable results for chloramination kinetic experiments with shorter storage time after quenching (i.e., maximum 3 days). However, it should be also noted that the use of Na_2SO_3 is only applicable for laboratory research investigations. In full-scale sampling programs, the residual concentrations can vary widely in distribution systems, therefore a large quenching agent concentration is used in the traditional USEPA methods (i.e., 100 mg/L NH_4Cl).

Overall, these results explain the main cause of the discrepancies for the reported HAA formation kinetics during chloramination in the literature. This discovery was published for the first time in the literature during the course of the project (Hong et al. 2005a,b).

EFFECT OF QUENCHING AGENT IN PRACTICAL APPLICATIONS

It should be noted that the results of the quenching agent experiments presented above were obtained using preformed chloramine. However, preformed chloramine is not commonly used in practical applications. The AWWA Disinfection Committee survey published in year 2000 showed that chlorine is added an average of 26 min. prior to ammonia addition in water treatment plants using chloramination in the US (American Water Works Association Committee Report 2000). Therefore, it was also decided to examine the impact of using NH_4Cl as "quenching agent" in practical applications. Another important motivation to examine the effect of quenching agent is the more stringent DBP regulations imposed by the USEPA under Stage 2 D/DBP rule. The Stage 2 D/DBP Rule is designed to reduce peak DBP concentrations in distribution systems, in part, by potentially changing D/DBP Rule compliance monitoring locations from those used in Stage 1. To identify potentially new monitoring locations, an Initial Distribution System Evaluation (IDSE)

Table 4.2
Concentrations of HAA species stored in the absence (0 mg/L) and presence of Na_2SO_3 (0.5–1.0 mg/L) for three days at 4°C

pH	Target (μg/L)	Na_2SO_3 (mg/L)	CAA (μg/L)	BAA (μg/L)	DCAA (μg/L)	BCAA (μg/L)	DBAA (μg/L)	TCAA (μg/L)	BDCAA (μg/L)	DBCAA (μg/L)	TBAA (μg/L)
6	5	0	5	5	5	5	5	5	5	6	5
	5	0.5	6	5	5	5	5	5	6	6	5
	5	1.0	5	5	5	5	5	5	6	6	5
	25	0	24	24	25	25	25	25	28	29	24
	25	0.5	23	24	25	25	25	25	26	29	23
	25	1.0	23	24	25	25	25	25	26	27	21
9	5	0	5	5	5	5	5	5	6	6	5
	5	0.5	5	5	5	5	5	5	5	6	5
	5	1.0	5	5	5	5	5	5	5	6	5
	25	0	25	25	26	26	25	26	30	33	25
	25	0.5	25	26	26	26	26	27	30	32	25
	25	1.0	25	26	26	27	27	27	31	33	26

study must be performed by each water utility. The purpose of the IDSE studies is to determine if new compliance monitoring sites are needed, which more accurately represent high concentrations of regulated THMs and HAA5. Therefore, to obtain reliable HAA results from IDSE studies for chloraminated systems, the suitability of using NH_4Cl as a quenching agent needs to be demonstrated, especially if falsely high HAA results are determined for sites sampled close to the treatment plant.

Four kinetics experiments were performed side-by-side by using SP DOM collected with the RO system (Figure 2.1) from the inlet of Spartanburg water treatment plant. This water was selected for testing because the DOM has a relatively low $SUVA_{254}$ value (~3 L/mg-m), as typically found in the treated drinking waters after conventional treatment processes. The SP DOM was diluted to a DOC of about 3 mg/L and chloraminated at pH 7 and 8.3 to simulate typical drinking water conditions.

Kinetics experiments were performed under four different chloramination scenarios to simulate possible practical chloramination scenarios:

- Case I: preformed chloramines,
- Case II: NH_3 immediately followed by chlorine addition,
- Case III: chlorine added to water 10 min. prior to NH_3 addition, and
- Case IV: chlorine added to water 30 min. prior to NH_3 addition.

In all cases, the goal was to maintain 2 mg/L total chloramines residual after 24-hr contact time, which represents typical residual concentrations in chloramination systems. The target times in the kinetics experiments (at the time of NH_3 addition (if applicable), 1, 2, 4, 8, 24, 48 and 72 hours) represent locations in a water treatment plant and/or distribution system with different residence times. For selected contact times (i.e., 4 and 24 hours), additional bottles were packed as duplicates. For each contact time, four bottles were prepared:

- Bottle #1 was quenched immediately with a stoichiometric amount of Na_2SO_3 to determine the true HAA concentration at the contact time. Although previous experiments have shown that there was no impact of 0.5 to 1.0 mg/L Na_2SO_3 residual on the stability of HAA during the two days of storage period, this bottle was stored at 4–6°C and extracted within eight hours after the Na_2SO_3 addition.
- Bottles #2 and #3 were quenched using 100 mg/L NH_4Cl, as required by USEPA Method 552.3. They were stored immediately in a refrigerator (4–6°C) for 2 and 7 days, respectively, before extraction. Comparison of the results for bottles #2 and #3 with that of bottle #1 showed the impact of the storage time and using NH_4Cl on the HAA determination at different reaction times (i.e., different locations in a distribution system).
- Bottle #4 was also quenched using NH_4Cl. However, it was stored initially in an ice chest with ice packs for one day, simulating the shipment of a sample to a laboratory, and subsequently 6 days (4–6°C) in a refrigerator before analysis. The result of bottle #4 was compared with those of bottles #1 and #3 to assess the impact of a typical sample handling procedure on HAA determination.

Experiments were performed at the ambient bromide levels (Br/DOC = 8 μg/mg) and pH 7 for all four cases. Since HAA formation is pH dependent, additional experiments were conducted at pH 8.3 that were selected based on the pH 7 results. The residual chloramines and pH were monitored during the experiments.

One important aspect of sample handling for DBP samples is the preservation of the samples at a low temperature (i.e., 4–6°C) before extraction. This is particularly important for chloramination samples quenched with NH_4Cl since the residual chloramine in the bottles was not destroyed. Without proper refrigeration, NH_2Cl is expected to continue to react with DOM and bromide at a faster rate as compared to refrigerated samples. In fact, warmer temperature may develop in ice chests during sample shipment. In this study, the HAA concentrations measured in bottles #3 and #4 were almost identical (Table 4.3). Statistical analyses showed that the results of HAA formation for bottles #3 and #4 were not different (p-value 0.33) for all four chloramination cases and two pHs (a total of 6 conditions and 54 samples). Therefore, properly keeping the samples in an ice chest with ice packs did not have any impact on the HAA concentrations. Since the results for bottles #3 and #4 were statistically the same, the results for bottle #4 are omitted from the discussion in the next section.

pH 7 Results

The results for pH 7 are shown from Figure 4.8 to Figure 4.11, and in a tabular format in Appendix D. For Case I (preformed chloramination) and Case II (addition of NH_3 immediately followed by HOCl addition) DCAA was the only HAA specie formed (Figure 4.8). The concentrations of DCAA in two cases were very similar, within a difference of 1–2 μg/L which is about the analytical variability of the measurements. This indicated that a good degree of mixing was achieved for Case II experiments. For the 1-hour samples, 3 μg/L of DCAA formation was observed for Case II and only 1 μg/L for Case I. It is possible that a small amount of initial HAA formation in Case II may be a result of HOCl reaction with very reactive DOM sites. Overall, the results show that there was no major difference in HAA formation between the preformed

Table 4.3
Comparison of HAA yields in bottles #3 and #4 during experiments

Time	HAA$_9$ (μg/L) @ pH 7								HAA$_9$ (μg/L) @ pH 8.3			
	Case I		Case II		Case III		Case IV		Case I		Case IV	
(hr)	#3	#4	#3	#4	#3	#4	#3	#4	#3	#4	#3	#4
1.0	10	11	11	11	37	38	47	45	8	8	35	31
2.0	11	11	12	11	38	39	46	46	8	8	33	32
4.0	10	11	12	11	37	39	47	46	8	8	33	34
4.0	11	11	11	13	37	39	46	46	8	8	34	34
8.0	12	11	12	12	38	39	47	45	8	9	33	34
24	13	13	13	14	39	40	46	48	8	8	33	34
24	13	14	14	14	39	40	48	46	8	8	34	34
48	14	15	15	15	41	40	47	48	9	9	37	35
72	15	16	16	16	39	40	47	49	9	9	38	34

chloramine addition and forming chloramine in-situ, as long as good mixing conditions were obtained during NH_3 and HOCl additions.

In both cases, DCAA formation was about 13 μg/L at 72-hr reaction time. The samples quenched with NH_4Cl showed higher DCAA formation and faster formation kinetics than the samples quenched with Na_2SO_3, especially at early reaction times. After one hour, formation of only 1 μg/L of DCAA was observed for the sample quenched with Na_2SO_3 in Case I, which was about 10% of the 72-hour yield. In contrast, DCAA concentrations in the 1-hr samples quenched using NH_4Cl were measured as 8 and 10 μg/L after two and seven days of storage time, respectively. These correspond to approximately 60–70% of the 72-hr yields. The difference between the two kinetic patterns (i.e., NH_4Cl vs. Na_2SO_3) decreased with increasing reaction times. For samples quenched with NH_4Cl, the bottles at the early reaction times had probably a small amount of HAA formed. However, since the NH_2Cl residual was not quenched as a result of NH_4Cl addition, it had two and seven more days (i.e., the storage time until the extraction) to react with DOM. On the other hand, for the case of Na_2SO_3, since the NH_2Cl residual was completely destroyed at the time of quenching agent addition, significantly lower amount of HAA formation was measured as compared to the samples quenched with NH_4Cl, especially at the early reaction times. The HAA concentrations (quenched with NH_4Cl) measured after the seven-day storage period were always slightly higher (1–2 μg/L) than those measured after two days of storage for both cases, indicating that the impact of storage period on the HAA concentration was not proportional to time. This is because the majority of the direct reactions between DOM and NH_2Cl was probably completed after two days of reaction time, and the low storage temperature was another reason for the small difference between the two conditions.

For Case III experiments, HOCl was contacted with DOM for 10 min. prior to NH_3 addition. The results are shown in Figure 4.9. Prechlorination, although it was only for 10 min., resulted in significantly higher HAA formation as compared to the first two cases. HAA yield was 37 μg/L at 72-hr for Case III, while it was only 13 μg/L for Cases I and II. These results confirm that chloramination, significantly reduces HAA formation, if HOCl and NH_3 are added simultaneously with good mixing conditions in water.

The speciation of HAA also changed as a result of the 10 min. prechlorination. Although DCAA was the only specie formed during performed or in-situ chloramination, formation of both DXAA and TXAA, including both chlorinated and a small amount of bromochlorinated species, was observed as a result of prechlorination (Figure 4.10). The formation of TXAA was due to prechlorination since chloramination does not form TXAA. After ammonia addition, the formation of TXAA remained constant at 17±1 μg/L over the 72-hr reaction period without any increase. In contrast, the formation of DXAA was the combination of 10 min. prechlorination and the 72-hr chloramination period since both disinfectants react with DOM to produce DXAA. Immediately before the ammonia addition, the DXAA concentration in the sample was 12 μg/L, and it increased gradually to 20 μg/L during 72-hr reaction period. Thus, an additional 8 μg/L of DXAA was formed due to chloramine reactions.

In terms of quenching agent effect, as expected, the maximum impact was observed at the early reaction times and the quenching agent effect decreased with increasing reaction time. The highest HAA concentration difference between Na_2SO_3 and NH_4Cl quenched samples were observed during the first two hours, which was 7 and 9 μg/L for two and seven-day storage periods, respectively. This difference was due to DXAA formation, providing evidence that chloramine in the bottles was continuing to react with DOM until the extraction time, because chloramines form only DXAA. As observed with the first two cases, five days longer storage time with NH_4Cl resulted in only slightly higher (1–3 μg/L) HAA concentrations. Although this difference was close to the variability in the analytical measurements, the seven-day samples consistently showed higher HAA yields than two day samples in this study.

In Case IV, the chloramination experiment was performed with 30 min. prechlorination followed by NH_3 addition (Figure 4.11). Similar to Case III, prechlorination resulted in a significant amount of HAA formation. The HAA formation during 72-hr reaction period was about 46 μg/L, which was 9 μg/L higher than Case III. Extending chlorination time for Case IV, at approximately same Cl_2/DOC ratio as Case III, resulted in mainly additional TXAA formation. The DXAA and TXAA concentrations before NH_3 addition were 13 and 17 μg/L for Case III (i.e., 10 min. prechlorination) and 15 and 23 μg/L for Case IV (i.e., 30 min. prechlorination), respectively. Addition of NH_3 completely shut down TXAA formation, while DXAA formation continued to occur (Figure 4.12). However, this was only noticeable in the samples quenched with Na_2SO_3. Since NH_4Cl did not quench chloramine, HAA formation continued to occur in the samples until extraction. Similar to Case III, the difference between Na_2SO_3 and NH_4Cl quenched samples was greater at the early reaction times (up to 10 μg/L) and decreased with reaction time.

Figures 4.13 and 4.14 show the positive errors in the HAA determination as mass concentration and the % error as a function of reaction time (or distribution system residence) for Case II, Case III and Case IV. In these figures, for each case the error as mass concentration was calculated by subtracting the HAA concentration quenched with Na_2SO_3 from that quenched with NH_4Cl (seven days storage) at each reaction time. The % positive error was calculated as the ratio of $[HAA_{NH4Cl}\text{-}HAA_{Na2SO3}]/HAA_{Na2SO3}$ for each reaction time. The errors in HAA determination result primarily from DXAA formation because DXAA was the only class of HAA formed during chloramination (Figure 4.15 and 4.16). The error trends were more apparent for DXAA than total HAA, indicating the impact of some variability in TXAA measurements for Case III and Case IV in the error analyses. Overall, the error analysis of the results obtained at pH 7 showed that for the samples within the first 8-hr residence time in a distribution system, it is possible to have up to a positive error of 10 μg/L for HAA by using NH_4Cl as the quenching agent and storing the samples in a refrigerator for seven days. The magnitude of the error diminished with increasing reaction time over eight hours.

Decay of NH_2Cl for these four cases is shown in Figure 4.17. For prechlorination cases, NH_2Cl decay was calculated based on NH_2Cl concentration in water immediately after NH_3 addition. A higher degree of NH_2Cl decay was observed for Cases I and II than for Cases III and IV, indicating that the initial prechlorination consumed some DOM reactive sites. Extended prechlorination period (from 10 to 30 min.) did not have a significant impact on the NH_2Cl decay.

The formation of DXAA in Case III and Case IV only caused by chloramination (total DXAA minus DXAA formed during prechlorination) were much less and the formation profiles were flatter than preformed chloramination (Case I), as shown in Figure 4.18. This indicated slower kinetics of DXAA formation after even 10 min. prechlorination. These observations also suggest that prechlorination consumed the DOM reactivity potential toward HAA formation, and chlorination and chloramination appear to attack the same DOM reactive sites.

Reckhow and Singer (1985) developed a generalized conceptual model for THM and HAA formation from fulvic acids during chlorination. They used a β-diketone moiety, (i.e., R'-CO-CH_2-CO-R) to represent the reactive site of DOM. The moiety will quickly become fully substituted with chlorine forming R'-CO-CCl_2-CO-R. Later, hydrolysis occurs rapidly, yielding a monoketone group, $CHCl_2$-CO-R. Then, DCAA formed if the R group is a hydroxyl group, otherwise, the structure will be further chlorinated to a trichloromethyl species CCl_3-CO-R, the precursor for TXAA and THM. They reported that, TXAA will be formed, if R is a readily oxidizable functional group, otherwise, THM will form. The absence of TXAA and THM formation during chloramination supported this model, assuming NH_2Cl is a weak oxidant and it cannot further chlorinate this $CHCl_2$-CO-R group into CCl_3-CO-R structure.

pH 8.3 Results

Since the pH 7 results for Case I and Case II were similar, the trends for Case III and Case IV were comparable, and Case IV represents an average chloramination condition in practice, Case I and Case IV were selected for testing at pH 8.3. For Case I (preformed chloramination) the quenching agent effect was observed (Figure 4.19). Only about 5 μg/L DCAA formed after 72-hr, which was much lower than at pH 7. The greatest impact of the quenching agent was observed at the earliest reaction period and the use of NH_4Cl resulted in 5 and 8 μg/L overestimation at the first hour for two-day and seven-day storage times, respectively, as compared to the samples quenched with Na_2SO_3. The effect of storage time on the HAA concentrations was not proportional to time. The difference between two-day and seven-day storage times remained within 2–3 μg/L, which was similar to the results observed for pH 7 conditions.

Case IV at pH 8.3 resulted in significant HAA formation as shown in Figure 4.20. At 72-hr reaction time, HAA concentration was 32 μg/L, which was much higher than for Case I. The HAA yield of Case IV was lower at pH 8.3 than at pH 7. As for pH 7, prechlorination at pH 8.3 resulted in the formation of TCAA and a small amount of brominated HAA (i.e., BCAA and BDCAA) which was not observed for Case I. Prior to NH_3 addition, 13 μg/L of TXAA formed, and after NH_3 addition it remained within 12–15 μg/L. The TXAA yield of prechlorination at pH 8.3 was 14 μg/L (Figure 4.21), which was lower than 23 μg/L at pH 7. This was consistent with the decreasing nature of TXAA formation with increasing pH during chlorination. In contrast, the formation of DXAA during prechlorination was 15 and 14 μg/L at pH 7 and 8.3, respectively, showing the pH independent nature of DXAA formation during chlorination. Similar to pH 7, the formation of DXAA in Case IV was a combined result of formation during prechlorination and chloramination. In terms of the quenching agent effect, the use NH_4Cl resulted in up to 5 and 8 μg/L higher amounts of HAA for two-day and seven-day storage, respectively, as compared to the Na_2SO_3 quenched samples.

As for pH 7 condition, an error analysis was also conducted for Case IV results at pH 8.3. Since there was no DXAA formation during the first eight hours of Case I, the same analysis was not performed for Case I. The results are presented in Figures 4.22 and 4.23 for total HAA, and Figures 4.24 and 4.25 for DXAA. The trends were in agreement with those obtained at pH 7 and were more apparent for DXAA than total HAA. Errors up to 8 μg/L HAA formation, primarily resulted from additional DXAA formation, were observed when NH_4Cl was used as the quenching agent. It was noted that for Case IV the errors in DXAA determination was maximum 65 and 58% for pH 7 and 8.3, respectively (Figure 4.19 and Figure 4.25). This suggests that the quenching agent effect decreases with pH, and the small difference observed here was probably due to only one pH unit difference.

The prechlorination and pH effects for Case I and Case IV are summarized in Figure 4.26. The data in this figure are for Na_2SO_3 quenched samples, therefore representing true HAA formation. It is evident that HAA formation can be significantly suppressed and kept well below the Stage 2 D/DBR limits with chloramination, if NH_3 is added to the water prior to HOCl addition or they are added simultaneously and good mixing conditions are maintained at the mixing locations of these chemicals. An increase of one and half pH units significantly decreased HAA formation. This shows that for utilities using chloramination, HAA formation can be controlled with a combination of pH adjustment, optimum prechlorination and in-situ chloramine formation. For example, the 72-hr HAA yield was 45 μg/L at pH 7 with 30 min. prechlorination while at pH 8.3 with chloramines formed in-situ, it was only 5 μg/L.

Comparison of Quenching Agents

It is evident that selecting a proper quenching agent is important for obtaining accurate HAA results during chloramination. The use of an improper quenching agent is likely to lead to positive errors. In order to examine different agents for chloramination samples, several potential quenching agents were tested side by side in a limited number of kinetics experiments since finding and validating a proper quenching agent for chloramination was not one of the objectives in this project. Case IV, 30 min. prechlorination, pH 7, DOC = 3.1 mg/L, Br/DOC = 100 μg/mg was selected to simulate a typical chloramination condition. The bromide level was also set at a relatively high concentration to check the stability of brominated HAA species. Five quenching approaches, stoichiometric amount of Na_2SO_3, NH_4Cl, sodium meta-arsenite ($NaAsO_2$), increasing pH to 10 using a borate acid buffer solution, and increasing pH to 10 followed by 100 mg/L ascorbic acid (i.e., Vitamin C) addition, were tested side by side using the same chloraminated samples. Samples quenched with a stoichiometric amount of Na_2SO_3 served as control samples and they were extracted within 8 hr after quenching agent addition. Samples quenched with other quenching agents were stored in the refrigerator for 7 and 14 days before extraction. Results are shown in Figure 4.27. Increasing the pH of samples to 10 gave the maximum error as compared to samples quenched with stoichiometric amount of Na_2SO_3. Repetition of the same experiment provided similar results. This was somewhat unexpected and the reason behind this observation is not known. Since the results were not promising this approach was not investigated further. The addition of ascorbic acid to high pH samples showed better results than pH adjustment alone, but almost as high as samples quenched with NH_4Cl. In addition, ascorbic acid almost completely destroyed brominated TXAA after seven days (Figure 4.27 C). It is expected the use of only 100 mg/L ascorbic acid shows the similar impact on HAA samples to the use of 100 mg/L of Na_2SO_3. They may preserve HAA6 but not the three brominated TXAA species. Among all methods tested, $NaAsO_2$ was found to be the best quenching agent, showing no effect on all nine HAA determination even after 14 days. Furthermore, the biocide capability of $NaAsO_2$ has been previously shown for distribution system samples (Brophy et al. 2000). Unfortunately, due its toxic nature, the use of $NaAsO_2$ in practical application poses various hazard and waste handling issues. More research is needed to identify and validate a proper quenching approach for chloraminated DBP samples in practical applications. The ideal method, although not always feasible, is the immediate extraction of a sample. If a short storage time is needed (1–2 days) and if the residual chloramine concentration can be quickly determined in a sample, the use of stoichiometric amount of Na_2SO_3 can be considered for a short storage period. It should be noted that Na_2SO_3 does not work as a biocide, thus degradation of HAA species may occur in biologically active samples.

SUMMARY OF IMPORTANT FINDINGS

Important findings from this phase of this study are as follows:

- It was found that the direct reaction of NH_2Cl with DOM is the major pathway for HAA formation during chloramination. Experiments performed in the presence (where NH_2Cl decomposition and $NHCl_2$ formation are significantly suppressed) and absence (where NH_2Cl decomposition can occur) of background ammonia indicated that a major amount (~80%) of DXAA was formed from the direct reaction(s) of NH_2Cl with DOM at pH 6. The remaining about 20% resulted from HOCl. Results showed that $NHCl_2$ did not form HAA at typical drinking water conditions. The contribution of HOCl to the overall DXAA formation decreased with increasing pH. The amount of DXAA formed from the direct reaction(s) significantly decreased with increasing pH. This indicates that even if NH_2Cl decomposition is minimized or NH_2Cl formed by mixing Cl_2 and NH_3 at the same injection location, an appreciable amount of DXAA may still form with decreasing pH values.
- It was shown that HAA formation kinetics during chloramination are not extremely fast, as previously reported in the literature. It was discovered and reported for the first time that the main reason for the different patterns reported for HAA formation kinetics in the literature is the use of NH_4Cl as the quenching agent recommended by USEPA Method 552.3. Since NH_4Cl does not quench the residual chloramine, the reactions of chloramine can continue until the samples are analyzed. As a result, falsely high HAA concentrations are measured at early reaction periods. The practical consequence of this experimental artifact is the possibility of overestimating the HAA formation for distribution system samples taken from locations close to the treatment plant. For one water tested in this study, it was found that for the samples within the first 8 hr of residence time in a distribution system at pH 7 and 8.3, a positive error up to 10 μg/L was observed in the measured HAA concentrations by using NH_4Cl as the quenching agent and storing the samples in a refrigerator for seven days. The magnitude of the error decreased with increasing residence time.

- In a limited number of experiments, five quenching approaches, stoichiometric amount of Na_2SO_3, NH_4Cl, sodium meta-arsenite ($NaAsO_2$), increasing pH to 10 using a borate acid buffer solution, and increasing pH to 10 followed by 100 mg/L ascorbic acid (i.e., Vitamin C) addition, were tested side by side using the same chloraminated sample. $NaAsO_2$ produced the best results, showing no effect on all nine HAA determinations even after 14 days of storage. Furthermore, the biocide capability of $NaAsO_2$ has been previously shown for distribution system samples. Unfortunately, due its toxic nature, the use of $NaAsO_2$ in practical application poses various hazard and waste handling issues. More research is needed to identify and validate a proper quenching approach for chloraminated DBP samples in practical applications. The ideal method, although may not be always feasible, is the immediate extraction of a sample. If a short storage time is needed (1–2 days) and if the residual chloramine concentration can be quickly determined in a sample, the use of stoichiometric amounts of Na_2SO_3 can be considered for a short storage period. It should be noted that Na_2SO_3 does not work as a biocide, thus degradation of HAA species may occur in biologically active samples. The use of excess Na_2SO_3 concentrations results in the decomposition of some brominated HAA species.
- The results also showed that chloramination is an effective way to control DBP formation, especially by minimizing the lag time between chlorine and NH_3 addition and maintaining pH above 7.5. Since HAA formation was found to occur mainly from the reaction of NH_2Cl with DOM, the enhanced removal of DOM can decrease HAA formation during chloramination.

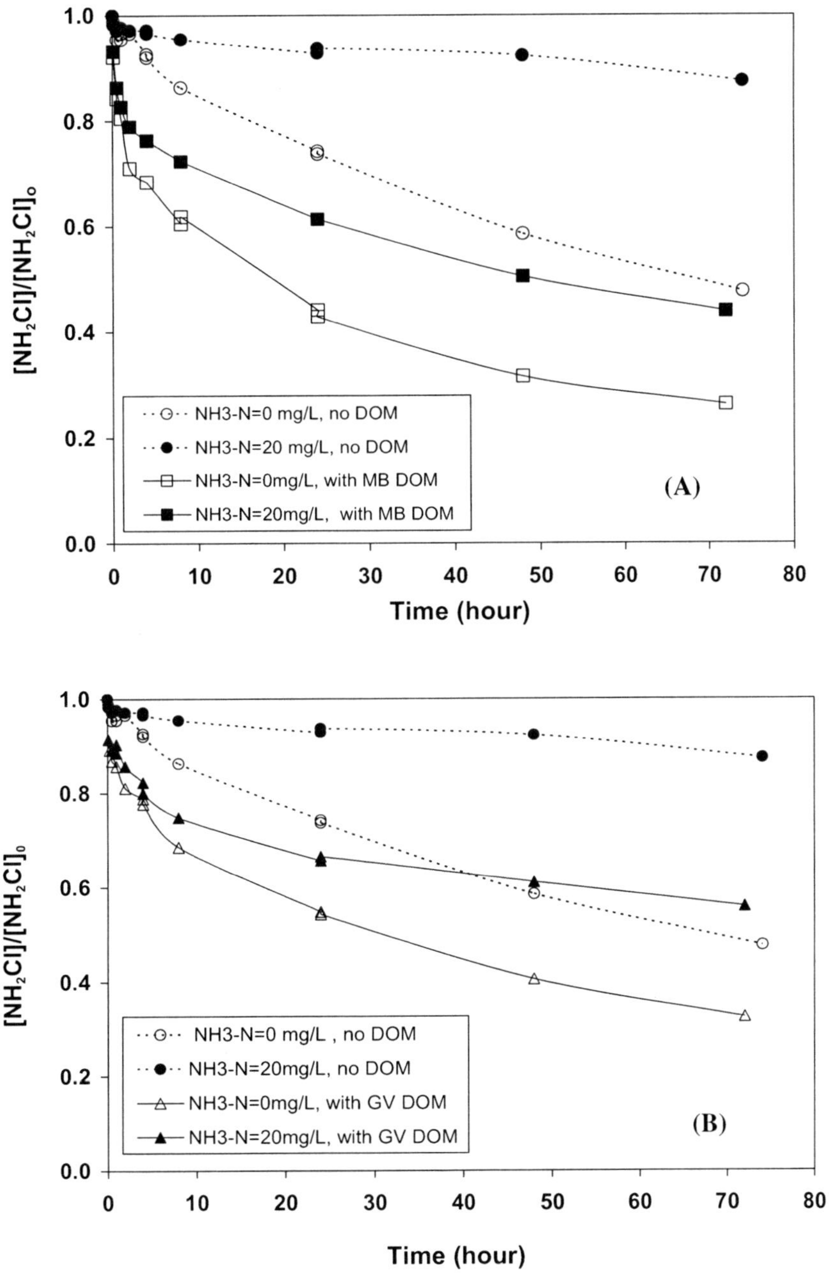

Figure 4.1 Decomposition of NH_2Cl in the absence and presence of DOM (7.5 mg/L DOC) with (NH_3 = 20 mg/L) and without addition of background ammonia at pH 6 for (A) Myrtle Beach and (B) Greenville DOM

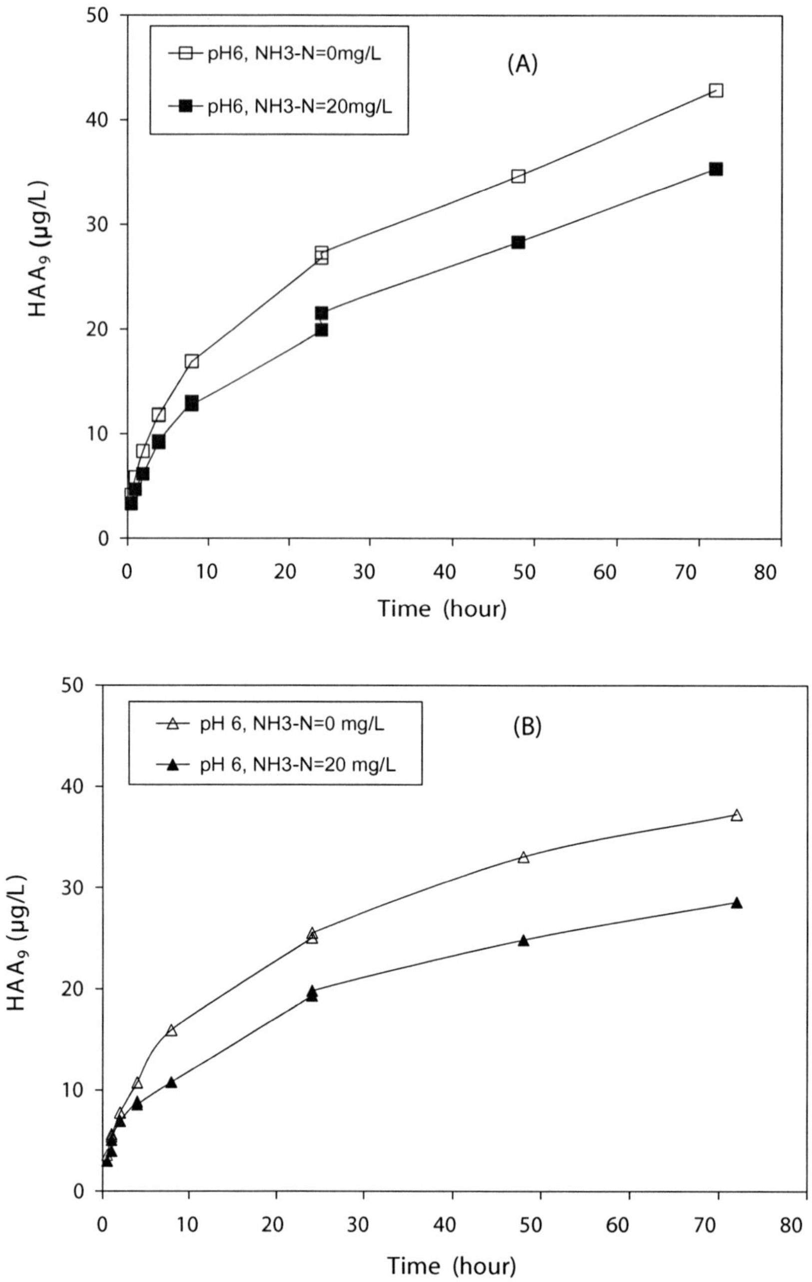

Figure 4.2 Formation of HAA with (NH_3 = 20 mg/L) and without addition of background ammonia at pH 6 for (A) Myrtle Beach and (B) Greenville DOM for DOC of 7.5 mg/L

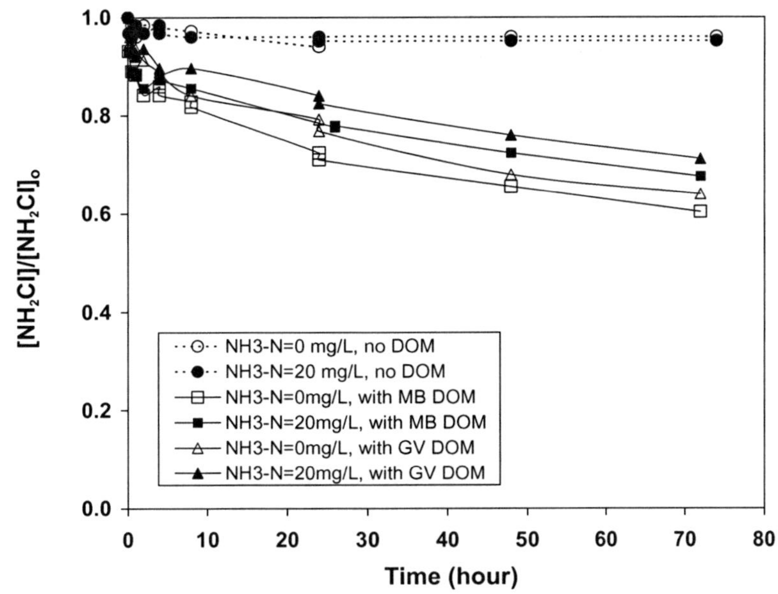

Figure 4.3 Decomposition of NH_2Cl in the absence and presence of DOM (7.5 mg/L DOC) with (NH_3 = 20 mg/L) and without addition of background ammonia for Myrtle Beach and Greenville DOM at pH 9

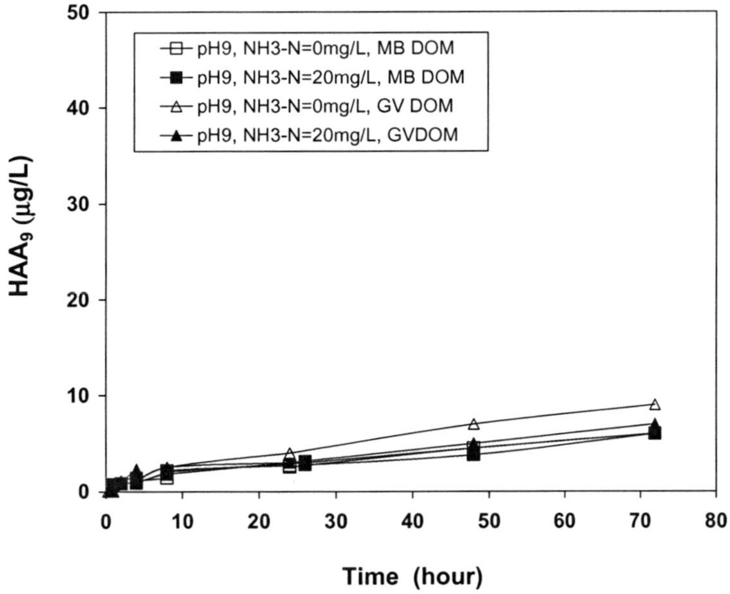

Figure 4.4 Formation of HAA with (NH_3 = 20 mg/L) and without addition of background ammonia in Myrtle Beach and Greenville DOM at pH 9

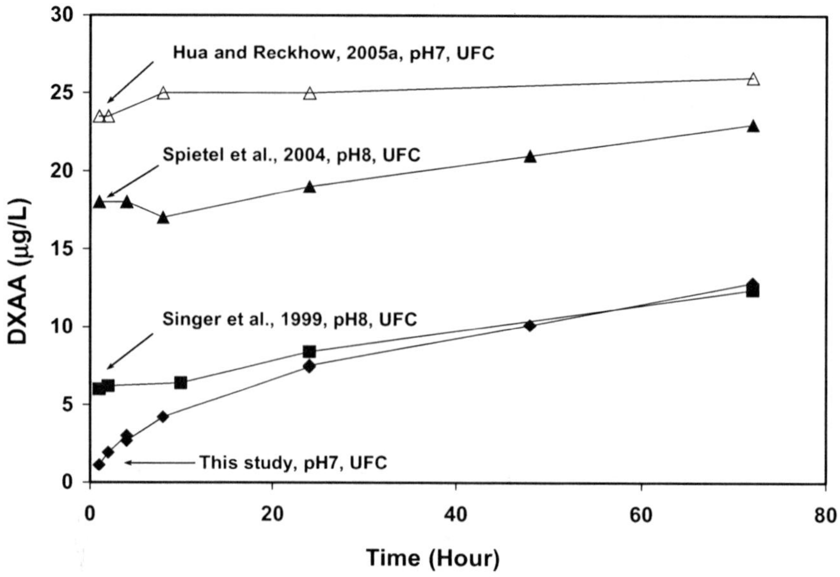

Figure 4.5 DXAA formation kinetics during chloramination, reported by different researchers

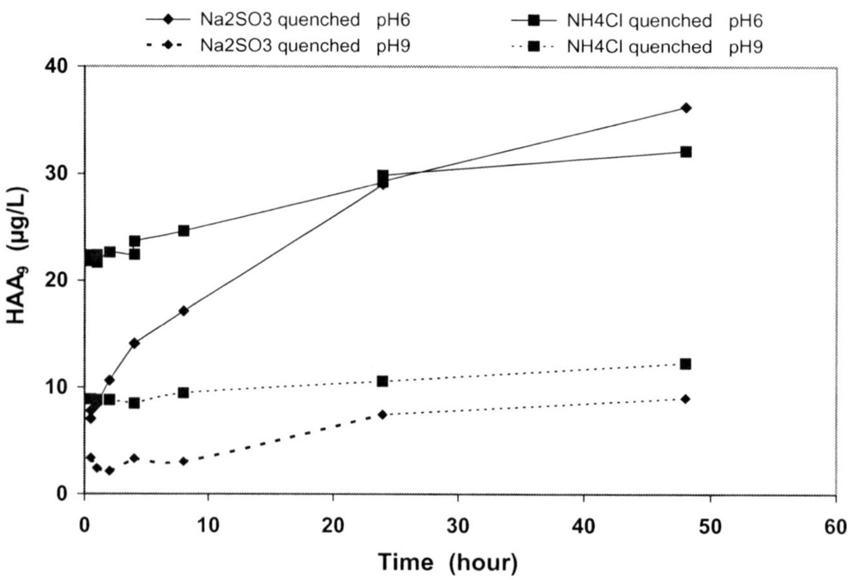

Figure 4.6 HAA formation kinetics determined at pH 6 and 9 using Na_2SO_3 or NH_4Cl as quenching agents

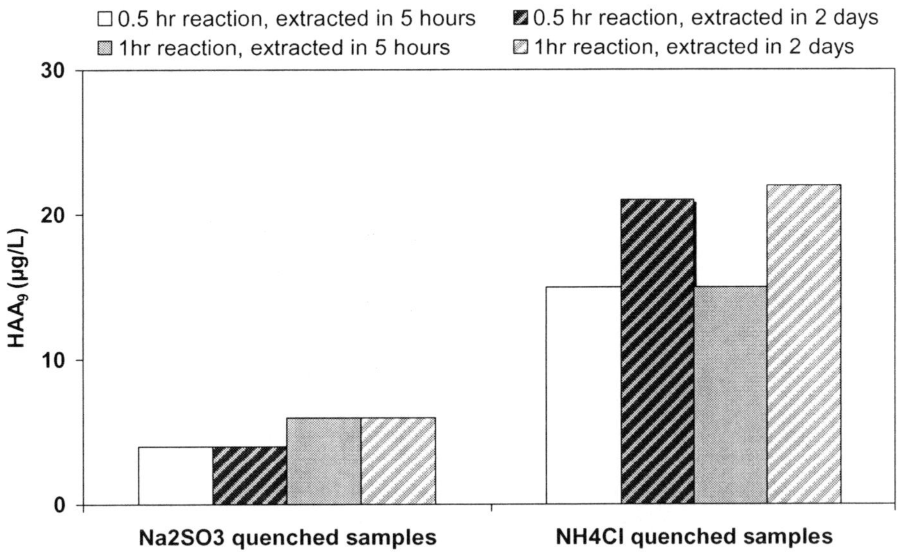

Figure 4.7 Effects of quenching agents and holding time on HAA formation

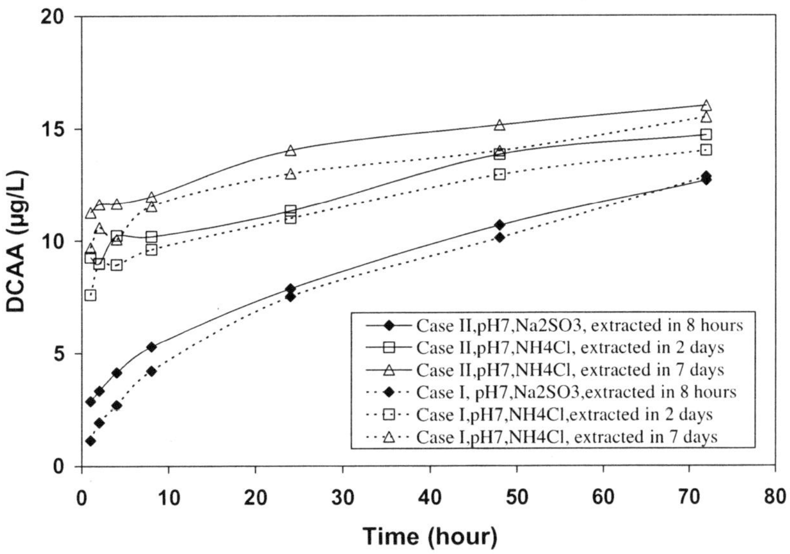

Figure 4.8 Effects of quenching agent and storage time on HAA formation for Cases I and II at pH 7

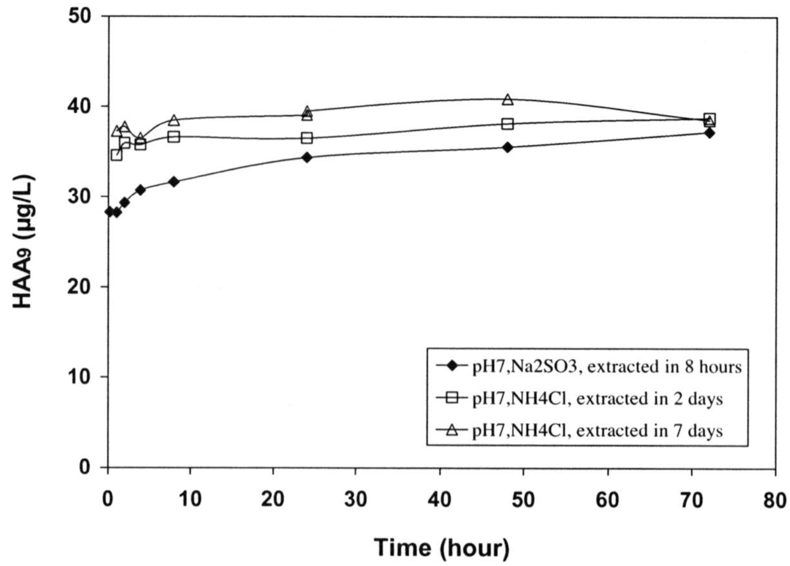

Figure 4.9 Effects of quenching agent and storage time on HAA formation for Case III (prechlorination for 10 minutes before ammonia addition) at pH 7. The time on the x-axis begins after the ammonia addition.

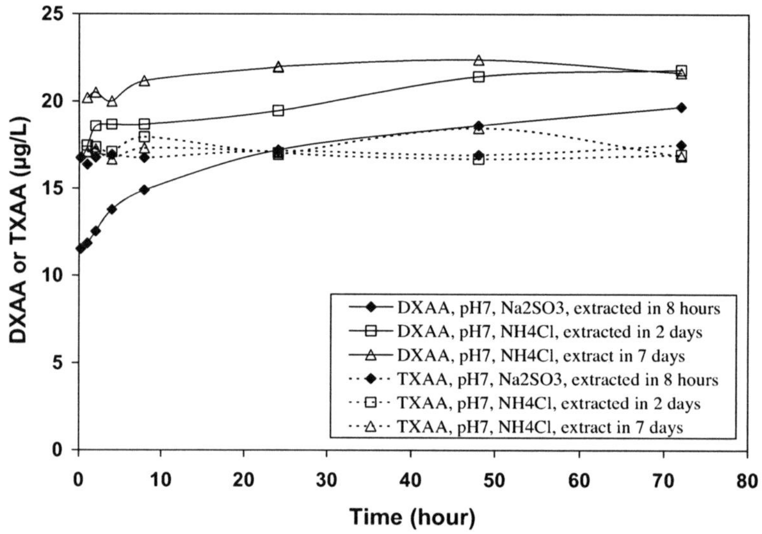

Figure 4.10 Effects of quenching agent and storage time on DXAA and TXAA formation for Case III (prechlorination for 10 minutes before ammonia addition) at pH 7. The time on the x-axis begins after the ammonia addition.

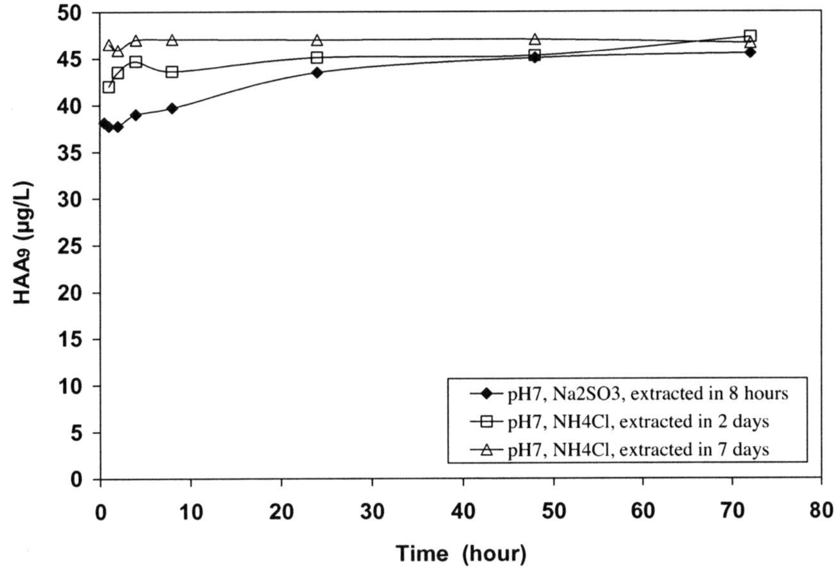

Figure 4.11 Effects of quenching agent and storage time on HAA formation for Case IV (prechlorination for 30 minutes before ammonia addition) at pH 7. The time on the x-axis begins after the ammonia addition.

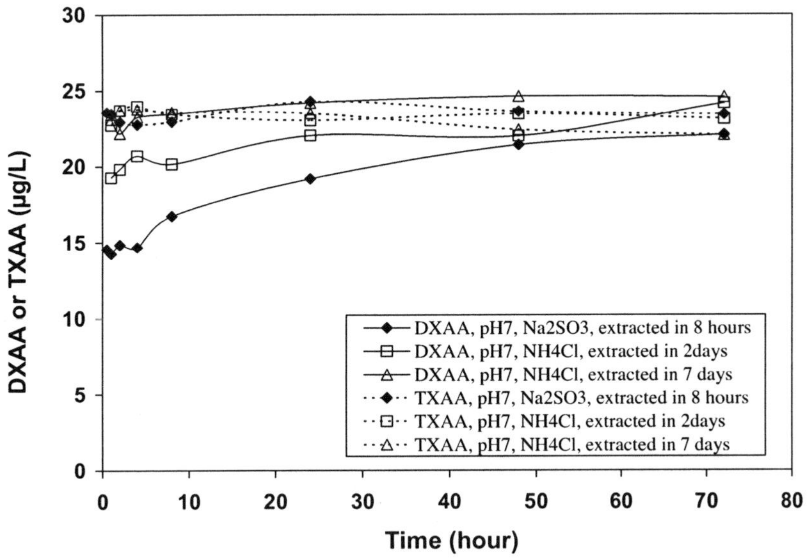

Figure 4.12 Effects of quenching agent and storage time on DXAA and TXAA formation for Case IV (prechlorination for 30 minutes before ammonia addition) at pH 7. The time on the x-axis begins after the ammonia addition.

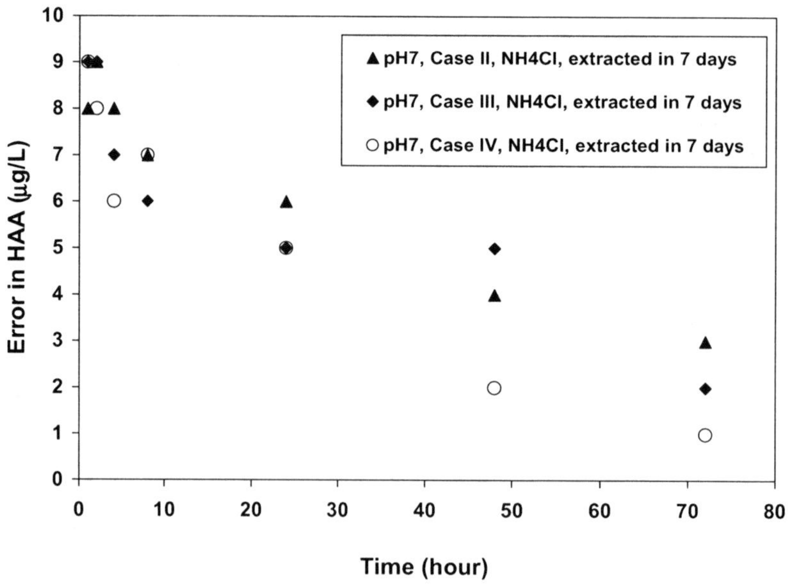

Figure 4.13 Error in HAA determination as mass concentration as a result of using NH_4Cl as quenching agent for three cases at pH 7

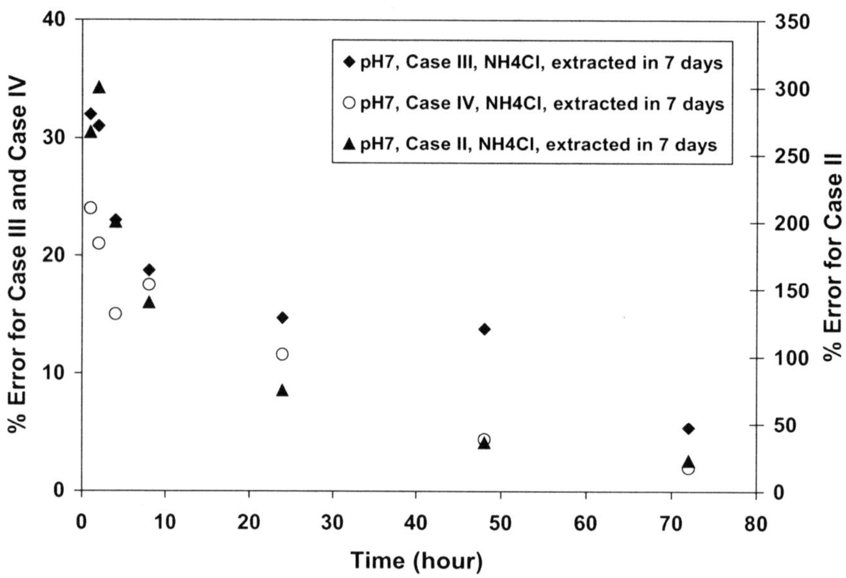

Figure 4.14 Percent error in HAA determination as a result of using NH_4Cl as quenching agent for three cases at pH 7

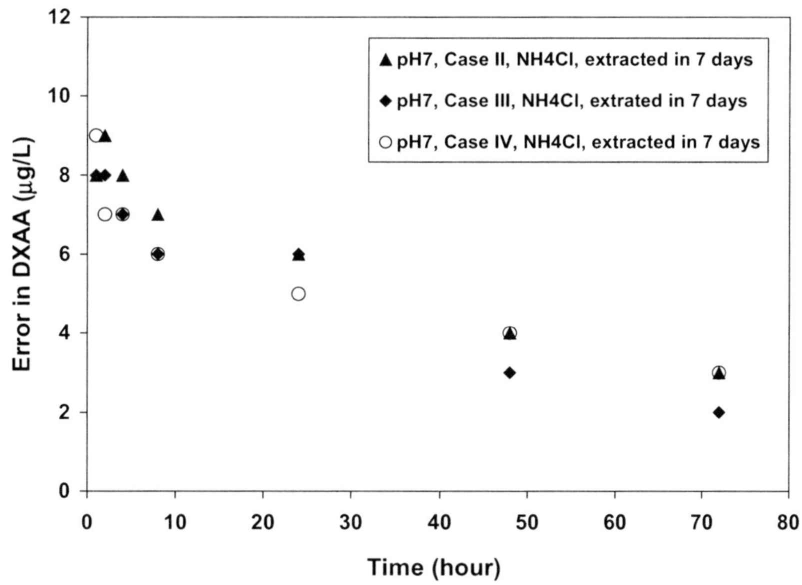

Figure 4.15 Error in DXAA determination as mass concentration as a result of using NH_4Cl as quenching agent for three cases at pH 7

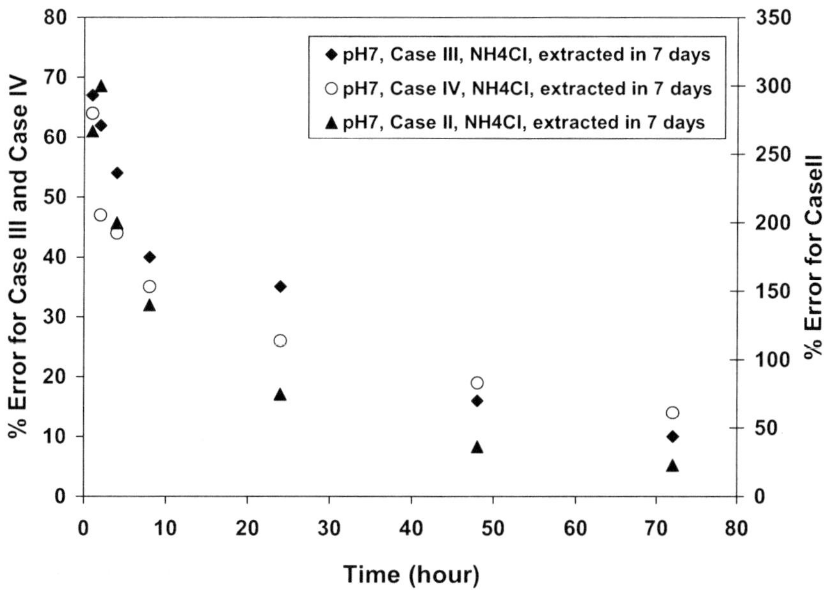

Figure 4.16 Percent error in DXAA determination as a result of using NH_4Cl as quenching agent for three cases at pH 7

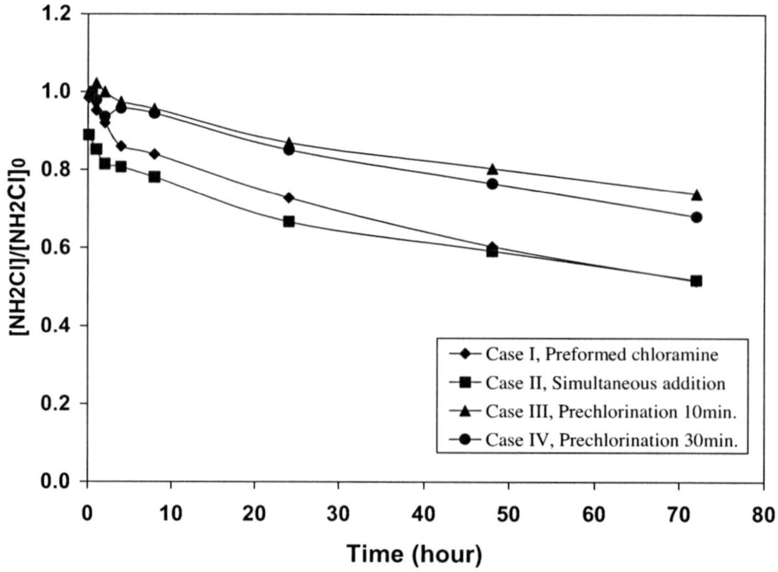

Figure 4.17 NH_2Cl decay during all four cases at pH 7

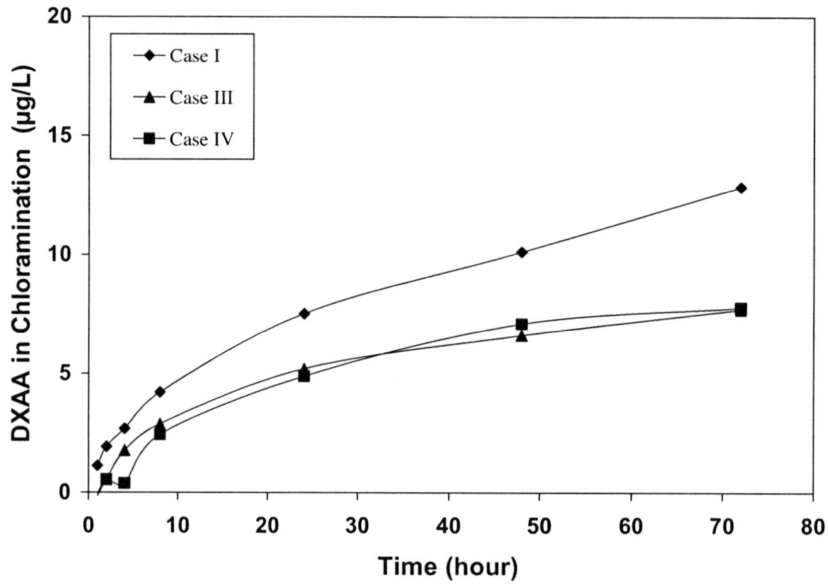

Figure 4.18 DXAA formation only in chloramination for Cases I, III, and IV

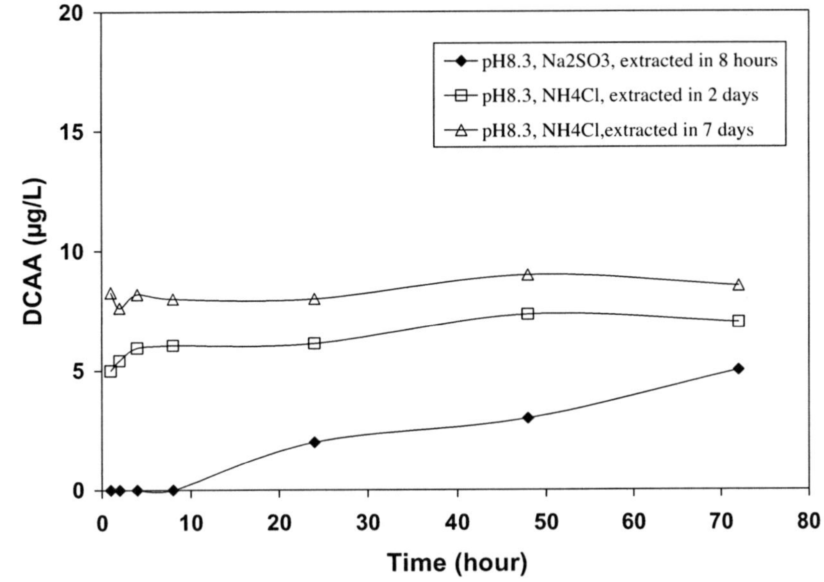

Figure 4.19 Effects of quenching agent and storage time on HAA formation for Case I (preformed chloramination) at pH 8.3

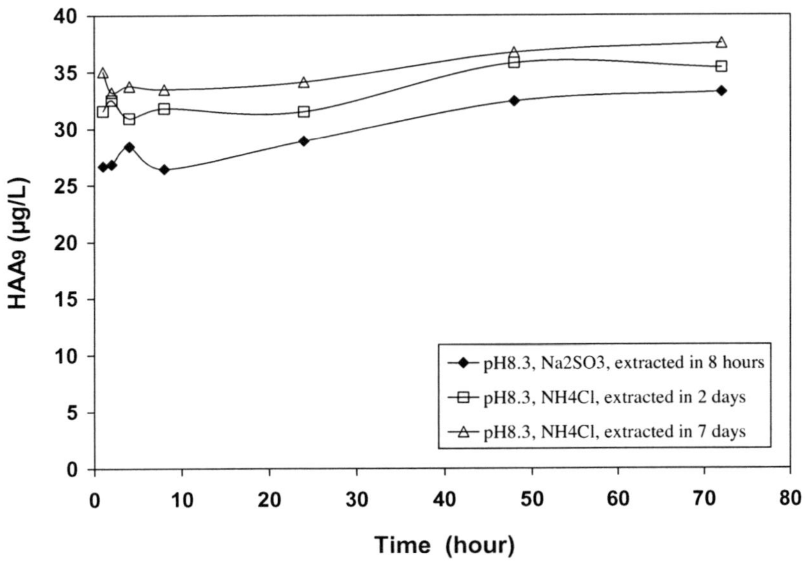

Figure 4.20 Effects of quenching agent and storage time on HAA formation for Case IV (prechlorination for 30 minutes before ammonia addition) at pH 8.3

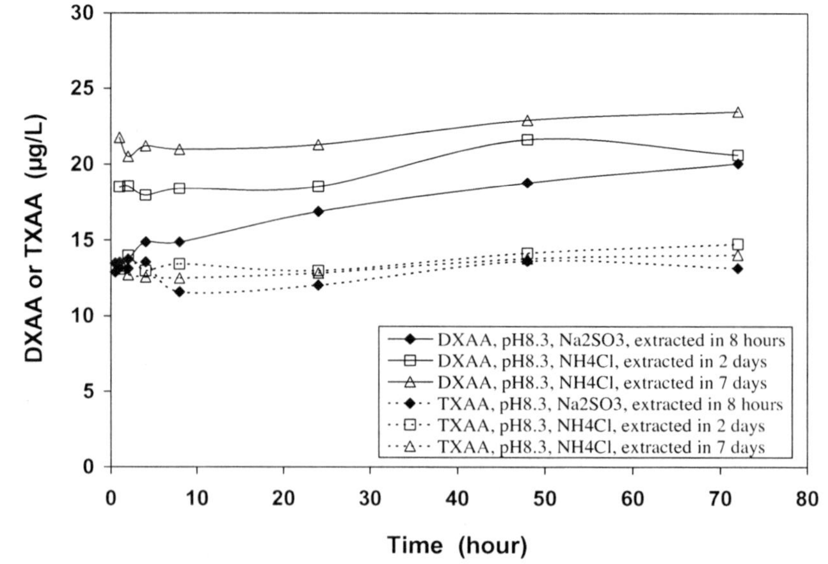

Figure 4.21 Effects of quenching agent and storage time on DXAA and TXAA formation for Case IV (prechlorination for 30 minutes before ammonia addition) at pH 8.3

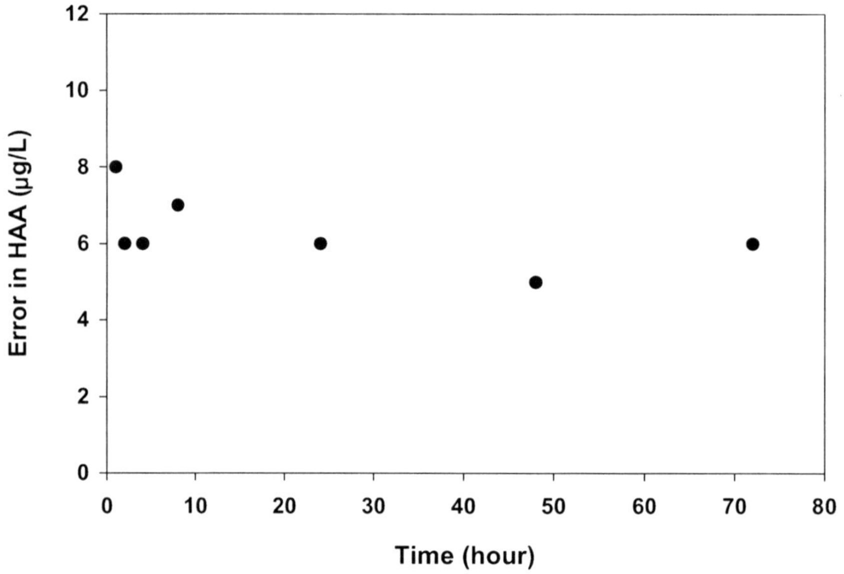

Figure 4.22 Error in HAA determination as mass concentration as a result of using NH_4Cl as quenching agent for Case IV at pH 8.3

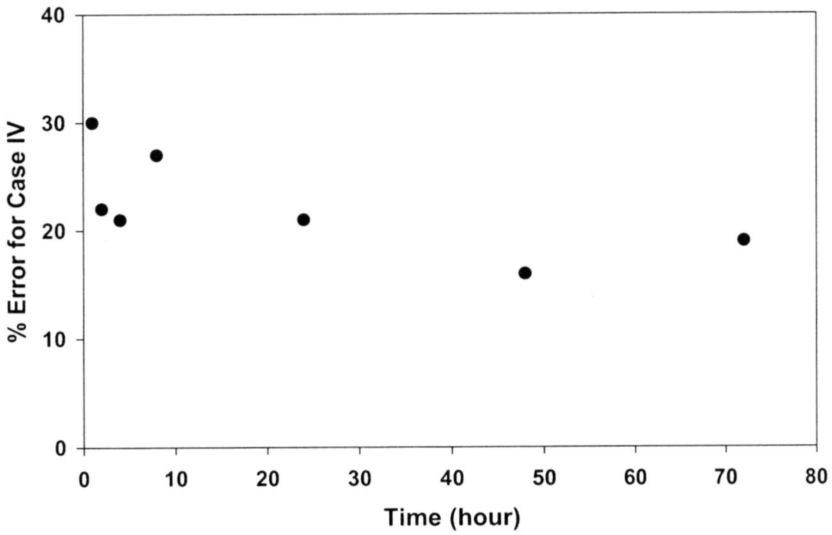

Figure 4.23 Percent error in HAA determination as a result of using NH_4Cl as quenching agent for Case IV at pH 8.3

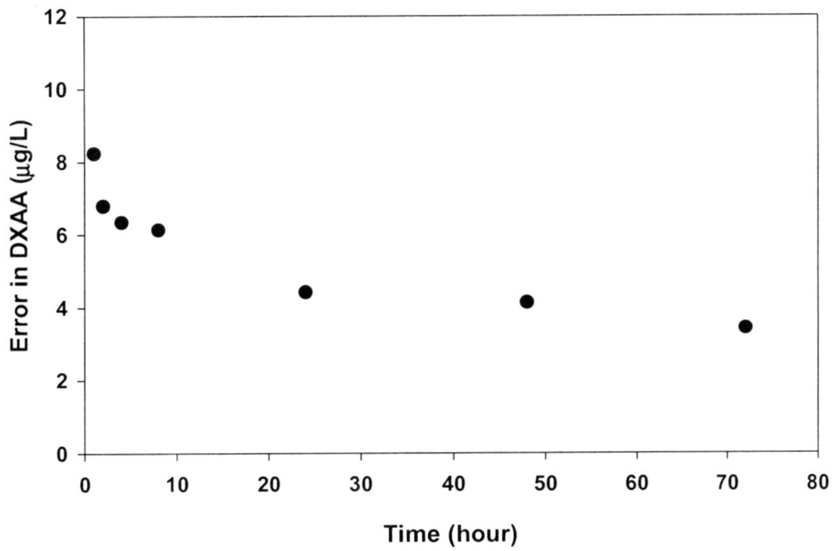

Figure 4.24 Error in DXAA determination as mass concentration as a result of using NH_4Cl as quenching agent for Case IV at pH 8.3

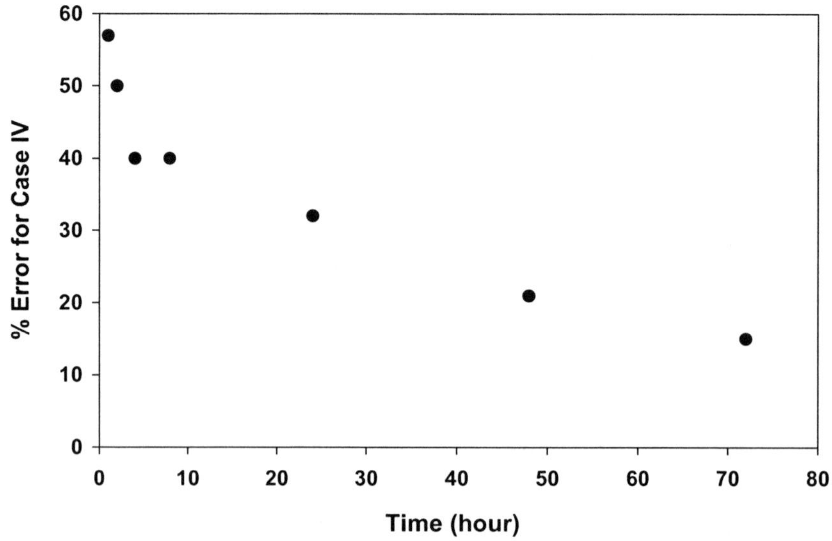

Figure 4.25 Percent error in DXAA determination as a result of using NH_4Cl as quenching agent for Case IV at pH 8.3

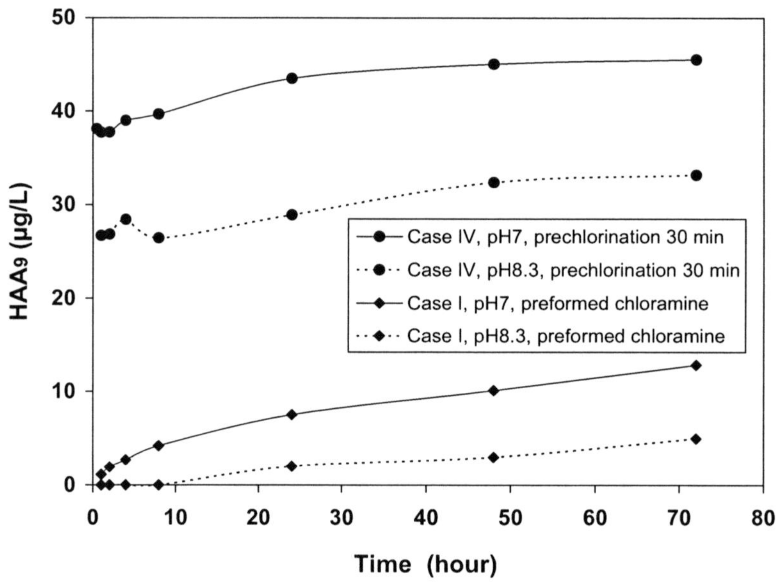

Figure 4.26 Effect of prechlorination and pH on HAA formation

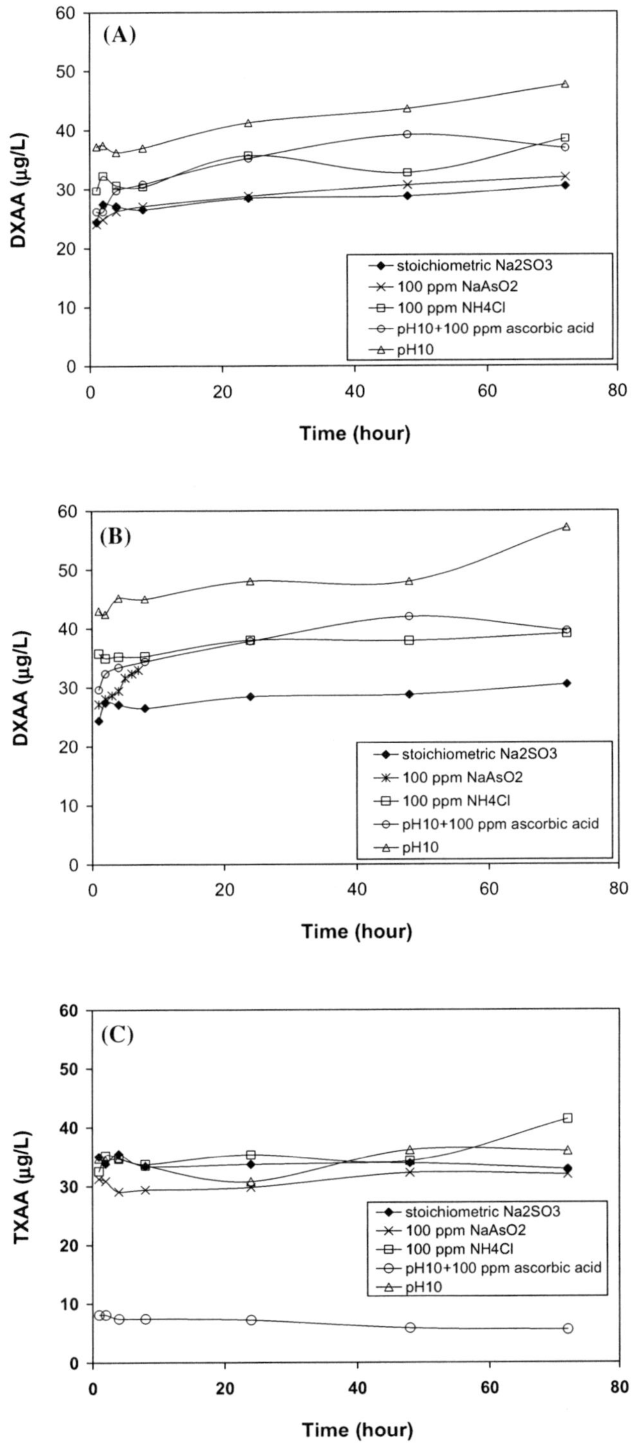

Figure 4.27 Quenching agents comparison: formation of (A) DXAA, 7 days, (B) DXAA, 14 days, and (C) TXAA, 7 days, for Spartanburg DOM at pH 7, Br/DOC = 100 µg/mg, and DOC = 3 mg/L

CHAPTER 5
HAA FORMATION FROM DOM AND DOM FRACTIONS DURING CHLORAMINATION

INTRODUCTION AND OBJECTIVE

Information about HAA formation from different DOM components during chloramination is relatively limited in the literature. Most of the HAA data are for 24-hr yields and thus information about HAA formation kinetics is very limited for both whole waters and DOM fractions. Furthermore, some of the available kinetic data are not reliable due to the type of quenching agent used in the experiments (Pope et al. 2002, Pope et al. 2003, Speitel et al. 2004), creating artifacts especially for the early formation kinetics, as discussed in the previous chapter and as reported elsewhere (Hong et. al. 2005a,b).

Duirk (2003) found a positive correlation between $SUVA_{280}$ and DCAA formation rate coefficients for three isolated DOM fractions from natural waters, but a commercial humic acid did not follow this trend. Speitel et al. (2004) reported that waters after conventional treatment processes always had lower DXAA reactivity (on a per organic carbon basis) than the source waters during chloramination. This indicated that HAA precursors were hydrophobic in character, since hydrophobic fractions are more amenable to coagulation than hydrophilic fractions (Collins et al. 1986, Singer and Harrington 1993). Recently, Hua and Reckhow (2005b) fractionated three surface waters. The chloramination result showed that the HPO and TPH fractions were more reactive in forming DXAA than the HPI fraction in a water with a high $SUVA_{254}$ value (4.4 L/mg-m). However, in two other low $SUVA_{254}$ waters (2.8 and 1.6 L/mg-m), the DXAA reactivity were similar for all three fractions at pH 7. In contrast to these reports, Hwang et al. (2000) studied Colorado River water, a very low $SUVA_{254}$ water (1–1.6 L/mg-m), and showed that the yield of DXAA from the hydrophilic acid and neutral fraction was almost three times higher than that from the hydrophobic acid fraction during chloramination. Furthermore, the hydrophilic fraction showed the highest reactivity toward THM and HAA formation during chlorination. They explained this difference using the model developed by Reckhow and Singer (1985), who suggested that high content of ketoacids (R'-CO-CH_2-COOH) in low $SUVA_{254}$ waters favored the formation of DXAA.

The objective of this phase of the study was to investigate the roles of various DOM components, pH and bromide concentration in the formation of HAA during chloramination. The approach was to systematically investigate the formation and speciation of HAA of different DOM components of the three surface waters with significantly different DOM characteristics, two with low $SUVA_{254}$ (Greenville (GV) = 2.6 mg/L-m, and Spartanburg (SP) = 3.2 mg/L-m) and one with high $SUVA_{254}$ (Myrtle Beach (MB) = 4.9 L/mg-m). DOM was isolated from each source using the RO system. The GV and MB DOMs were subsequently separated into three fractions, HPO, TPH and HPI, as described in Chapter 2. The general water quality characteristics at the source waters are provided in Table 3.1. The RO isolation enabled the chloramination experiments and DOM fractionation of the two source waters to be done at the same organic carbon concentration. The formation of HAA in the RO isolates (i.e., representing the whole water) and three fractions obtained from two of the three RO isolates were examined side by side. The roles of pH and bromide in HAA formation were studied at typical water treatment conditions for their practical significance. Furthermore, experiments were conducted to examine HAA formation kinetics of different DOM fractions and RO isolates as a function of pH and bromide concentrations.

HAA FORMATION OF RO ISOLATES

Speciation of HAA

DXAA in both RO isolates accounted for at least 80% of HAA (as mass or molar concentrations) formation after 72 hr for all bromide concentrations at pH 6 and pH 7.5 (Figure 5.1). At pH 6, the increase of Br/DOC ratio from ambient to intermediate level increased brominated DXAA formation, while decreasing MCAA formation. The further increase in bromide level increased MBAA and brominated TXAA formation, resulting in a decrease in the DXAA percentage. At pH 7.5, no MBAA and brominated TXAA were formed, and the DXAA percentage increased with bromide concentration. At pH 9, DCAA was the only HAA specie formed at all bromide concentrations (data not shown). It appeared that the distribution of MXAA, DXAA and TXAA was not sensitive to bromide, an observation similar to that for chlorination as reported by Cowman and Singer (1996).

In previous research using preformed chloramines, Spietel et al. (2004) reported that 72-hr DXAA formation represented at least 70% of HAA formation at pH 7 and 9 for different natural waters. MXAA formation was also observed but TXAA was, in general, negligible. Diehl et al. (2000) reported that at pH 6, 8 and 10, DXAA constituted typically more than 80% of the HAA6 monitored in their study. MCAA formation was observed at pH 6 and 8 in some waters, while formation of TCAA and MBAA was negligible at all three pH conditions. Cowman and Singer (1996) reported that DXAA, MXAA and TXAA constituted about 65, 30 and 5% of HAA formed as a result of chloramination of a DOM extract ($SUVA_{254}$ = 4.2 L/mg.m) at pH 8.

The results obtained in this study and in previous investigations with preformed chloramines indicate that at least 80% of HAA to be formed from chloramines(s) in natural waters will be in the DXAA form, and the contribution of DXAA may reach and exceed over 90% in some cases. A small amount of MXAA formation may also be observed, while formation of TXAA is usually negligible.

The results of HAA formation kinetics experiments for GV, SP and MB RO isolates are provided from Figures 5.2 to 5.4. The results are presented in a series of kinetics figures: NH_2Cl decay patterns followed by DXAA formation kinetics, and bromine incorporation factors. The complete set of data obtained during these experiments is presented in a tabular format in Appendix E.

Effect of pH

The results of NH_2Cl decay during the kinetics experiments are provided in Figure 5.2. As expected, the NH_2Cl decay rate increased with decreasing pH. At pH 6, NH_2Cl decay increased significantly with increasing bromide concentration, while no major impact of bromide concentration was observed on the NH_2Cl decay at pH 9. Only a small increase in NH_2Cl consumption was observed at the highest bromide concentration of pH 7.5 solution. Formation of $NHCl_2$ was only observed at pH 6.

The dependency of DXAA formation on pH was consistent with that of NH_2Cl decay. DXAA formation decreased with increasing pH at all three bromide concentrations for all three

Table 5.1
Comparison of HAA formation patterns as a function of pH during chlorination and chloramination

HAA	Chlorination pH 6	Chlorination pH 7.5	Chlorination pH 9	Chloramination* pH 6	Chloramination* pH 7.5	Chloramination* pH 9
MXAA	Essentially the same			Decreases with increasing pH		
DXAA	Essentially the same			Decreases with increasing pH		
TXAA	Decreases with increasing pH			Formation usually negligible, may occur at low pH		

* Using preformed chloramines.

waters (Figure 5.3). The formation of DXAA was very sensitive to pH. Figure 5.4 shows DXAA formation after the 72-hr chloramination period at three pHs. It is evident that lower pH favors HAA formation dramatically, especially for MB DOM having a high $SUVA_{254}$. The pH effect was not proportional to the change in pH. When pH was increased from 6 to 7.5, DXAA formation decreased by 18, 11 and 22 μg/L for GV, SP and MB DOM, respectively, while it only dropped 7, 10 and 9 μg/L when pH was increased another 1.5 unit from 7.5 to 9.

The results at pH 7.5 and pH 9 were compared with results reported by Duirk (2003) and Speitel et al. (2004). They both reported the same dependency of DXAA formation on pH, but with significantly different kinetics patterns. Duirk (2003) reported an increasing trend of DXAA formation with time up to 168 hr. This trend was, in general, comparable to those observed in this study, but only one data point was collected at reaction times less than 24 hour. On the other hand, Speitel et al. (2004) investigated HAA formation kinetics at early reaction periods in more detail. They reported very fast kinetics and similar patterns for HAA formation at pH 7 and pH 9, in which the formation of about 80–90% of the 24-hr HAA yields was observed within the first 0.5 to 1 hr. The different kinetic patterns observed by Speitel and co-workers as compared to those obtained in this study was attributed to the use of NH_4Cl as a quenching agent in their experiments, as previously discussed in Chapter 4.

It is clear from these results that water utilities intending to decrease HAA formation during chloramination should maintain pH above 7.5. It was shown that the formation of DXAA, MXAA and TXAA depend on pH during chloramination. This trend was different from those observed during chlorination (Table 5.1); suggesting that chloramination had its own specific reaction mechanism(s) of HAA formation.

Effect of Bromide

To assess the extent of bromine substitution in HAA, the bromine incorporation factor (BIF) was calculated:

$$BIF = HAABr6\ (\mu mol/L)/HAA9\ (\mu mol/L)$$

where HAABr6 is the sum of the molar concentrations of bromine incorporated in the six brominated HAA species. HAA9 represents the sum of molar concentrations of all nine HAA species. Therefore,

$$HAABr6 = (MBAA*1) + (DBAA*2) + (TBAA*3) + (BCAA*1) + (BDCAA*1) + (DBCAA*2)$$

During chloramination, the presence of bromide accelerated HAA formation and shifted HAA speciation to more brominated species (Figure 5.5). This observation was similar to that observed for chlorination, but to a much lesser extent. The effect of bromide depended on pH. At pH 6, DXAA formation and NH_2Cl decomposition increased with increasing bromide concentration (Figure 5.2 and Figure 5.3). As expected, the DXAA speciation shifted toward brominated species. Figure 5.6 (A) shows the change in distribution of DXAA species, (DCAA, BCAA and DBAA) with bromide concentration after 72-hr reaction time at pH 6. The DCAA formation decreased significantly with increasing bromide concentration to the intermediate level (Br/DOC = 150 μg/mg), while the DBAA and BCAA increased dramatically. However, the further increase of bromide concentration did not increase the contribution of BCAA to DXAA even though its formation increased with bromide concentration. This was because high bromide conditions favored DBAA formation over BCAA. At the highest bromide concentration after 72 hr, the mole percentage of DBAA was about 50%, while BCAA and DCAA accounted for about 30% and 20%, respectively. It was noted that the distribution of DXAA species was almost independent of DOM type at pH 6 in spite of the significantly different $SUVA_{254}$ value of the three DOMs. The presence of bromide resulted in the formation of trace amounts of brominated TXAA only at the highest Br/DOC ratios for all waters. At the ambient bromide levels, MCAA was the only MXAA species formed, and its formation decreased with increased bromide concentration. At the highest Br/DOC ratio, MBAA was the only MXAA formed.

At pH 7.5, the impact of bromide on DXAA formation and NH_2Cl decomposition was much less than at pH 6 (Figure 5.2 and Figure 5.3). It was essentially negligible at pH 9. At pH 7.5, the major HAA specie was DCAA even at the highest bromide concentration, accounting for 60–70% of total DXAA. Formation of DBAA was not observed until at the highest bromide concentration and its contribution was no more than 10% even after 72 hour reaction time, indicating that DBAA formation was more sensitive to pH than BCAA and DCAA when compared with the results at pH 6 (Figure 5.6). These results show that one additional advantage of having operational pH of 7.5 or above for chloramination was to reduce the formation of brominated HAA species.

Bromine incorporation was higher and occurred more rapidly at pH 6 than pH 7.5. The BIF increased significantly with increasing bromide concentration for both DOM (Figure 5.5). No brominated specie was formed at pH 9. At pH 9, NH_2Cl hydrolysis is unlikely to occur and produce appreciable amount of HOCl and consequently HOBr. As a result, NHBrCl formation from the NH_2Cl decomposition pathway (Reactions 2-5 to 2-7 in Table 1.2) is suppressed completely. Since formation of NH_3Cl^+ is an acid-catalyzed reaction (Trofe et al. 1980), it is also very unlikely for NHBrCl to form through the NH_3Cl^+ pathway (Reaction 2.1 in Table 1.2) at pH 9.

At pH 9, the NH_2Cl decay pattern reported by Duirk (2003) was similar to the results obtained in this study. In Duirk's study there was no effect of bromide on the NH_2Cl decay with the bromide concentration of about 0.1–5 mg/L (0.3–2.0 mg/L range used in this study), under the NH_2Cl concentration of 0.05 mM (similar to this study) and the DOM concentration 3.3 mg DOC/L. One noticeable difference at pH 9 between the results obtained in this study and those of Duirk (2003) and Speitel et al. (2004) was the formation of brominated HAA with increasing bromide concentration. In this study, formation of brominated HAAs was not observed at pH 9 in the three RO isolates. Duirk (2003) performed chloramination experiments with three DOM solutions of different $SUVA_{254}$ values at Br/DOC ratios of 32, 315 and 1515 μg/mg for seven days. He reported that no brominated species was observed at bromide levels < 350 μg/mg, while the formation of brominated species occurred (BIF about 0.5) at the extremely high Br/DOC ratio of 1515 μg/mg. It appears that although high pH suppresses brominated HAA formation; extremely high bromide concentrations may still result in brominated HAA formation after long reaction periods (e.g., seven days). In contrast to this study and Duirk (2003), Speitel et al. (2004) reported small amounts of BCAA and DBAA (each about 2–4 μg/L) formation for the Lake Austin source water ($SUVA_{254}$ = 2.5 L/mg-m, Br/DOC = 45 μg/mg, DOC = 3.74 mg/L). However, after addition of 0.5 mg/L of bromide (Br/DOC = 180 μg/mg), the authors reported significantly higher brominated HAA formation (BCAA about 3–5 μg/L and DBAA about 9–15 μg/L). This was not expected even though the experiments were performed under different temperature and Cl_2/N ratios. For another source water from Metedeconk River with a similar DOC concentration ($SUVA_{254}$ = 3.48 L/mg-m, Br/DOC = 9 μg/mg), the bromide effect was less significant, and the increase of BCAA and DBAA formation was only 2–3 μg/L (Speitel et al. 2004). The reason of the unusual bromide effect for Lake Austin at pH 9 was not clear. Besides these trends, Speitel et al. (2004) reported that bromine incorporation factor increased with decreasing pH in the waters tested, which was similar to this study. Moreover, the bromine incorporation in their study occurred almost instantaneously during the first 0.5 hr of reaction period. Again, the different BIF kinetic patterns observed by Speitel and co-workers as compared to this study are attributed to the use of NH_4Cl as a quenching agent in their experiments.

Effect of DOM Characteristics

The characteristics of DOM are expected to play a role in the formation and speciation of DBPs. In this study, experiments were performed with two relatively low-SUVA (GV and SP) and a high-SUVA (MB) water at identical chloramination conditions. The two low-SUVA DOM isolates showed relatively similar trends probably due to their close SUVA values. At pH 6, the formation of DXAA in MB DOM was about 10–30% higher (on a mass and mol basis) than in GV and SP DOM at three bromide levels (Figure 5.3). Since the DOC concentrations of the experiments were almost identical, these results indicate that a higher degree of DXAA reactivity was exhibited by the high-$SUVA_{254}$ DOM than the low-$SUVA_{254}$ DOM. However, the difference between the DXAA yields was rather small in comparison to the significantly higher $SUVA_{254}$ value of MB than GV and SP waters. The NH_2Cl decay patterns at pH 6 were consistent with the DXAA formation trends; more NH_2Cl consumption was measured in MB than GV and SP DOM (Figure 5.2). This was in agreement with the previous reports that NH_2Cl decomposition increased with the $SUVA_{254}$ of the DOM (Valentine et al. 1998). In terms of bromine effect, relatively similar bromine incorporation factors and trends were obtained for three DOM at pH 6 in spite of significantly different SUVA values (Figure 5.5). At the highest Br/DOC ratio, the GV and SP DOM had a slightly higher bromine incorporation factor than the MB DOM.

At pH 7.5 and 9.0, the impact of DOM aromaticity (i.e., $SUVA_{254}$) on NH_2Cl decay was negligible at each bromide condition (Figure 5.2). The DXAA formation patterns were consistent with their NH_2Cl decay patterns (Figure 5.3). All three DOM showed relatively similar DXAA yields. However, it was also noted that at pH 7.5, SP DOM showed significantly higher bromine incorporation than the other two DOM. Speitel et al. (2004) studied DXAA formation for two surface waters, Lake Austin water, $SUVA_{254}$ = 2.49 L/mg-m and DOC = 3.74 mg/L, and Metedeconk River water, $SUVA_{254}$ = 3.48 L/mg-m and DOC = 3.05 mg/L, at pH 7 and 9. The normalized DXAA yields with respect to DOC were the same for both waters at each pH in spite of one unit

SUVA difference between the waters. In contrast, Duirk (2003) reported almost three times higher DXAA yield from Upper Peninsula water ($SUVA_{254}$ = 3.05 L/mg-m) than from Iowa River water ($SUVA_{254}$ = 2.78 L/mg-m) at pH 8.3, although both had similar $SUVA_{254}$ values. Both waters had the same DOC concentration and NH_2Cl dose. This suggested the existence of highly reactive sites which were not reflected by the $SUVA_{254}$ of the Upper Peninsula water. No brominated HAA species was formed at pH 9. Bromine incorporation at pH 7.5 was significantly smaller compared to pH 6 and relatively similar for both DOMs at a Br/DOC ratio of 150 μg/mg (Figure 5.5).

Overall, these results show that the impact of DOM and bromide were significantly less pronounced with increasing pH, indicating that pH is a more important factor controlling the extent of DXAA formation and speciation than bromide and $SUVA_{254}$ of DOM under typical drinking water conditions.

HAA FORMATION OF DOM FRACTIONS

To investigate the role of DOM components on HAA formation, the RO isolates of GV and MB waters were separated into three different fractions (HPO, TPH and HPI) according to their hydrophobicity (Chapter 2 and Chapter 3). For all fractions, the experiments were performed at three Br/DOC levels (i.e., ambient (50 μg/mg), intermediate (130 μg/mg) and high (350 μg/mg)). The results for each fraction are presented in a series of figures: NH_2Cl decay patterns followed by DXAA formation kinetics and bromine incorporation factors (Figures 5.7 to Figure 5.15). The normalized NH_2Cl consumption and the normalized $NHCl_2$ formation during the 72-hr reaction period are provided in Table 5.2 and Table 5.3. The 72-hr HAA formation of the fractions for both waters is summarized in Table 5.4 and Table 5.5. The data collected for the results presented in this section is provided in a tabular format in Appendices F and G. The results are discussed below from pH 9 to 6 conditions, as the complexity of chloramination chemistry increases with decreasing pH, especially in the presence of bromide.

pH 9 Results

At pH 9, NH_2Cl was relatively stable (Figure 5.7A and Figure 5.8A). Similar NH_2Cl decay patterns and amounts were observed for both waters regardless of fraction type and bromide concentration (Table 5.2, Figure 5.7 and Figure 5.8). $NHCl_2$ was not detectable in any sample. It is unlikely that appreciable amounts of HOCl, and thus HOBr, form as a result of NH_2Cl hydrolysis at this pH. NHBrCl is the major bromo(chloro)amine specie that is expected be present. However, since formation of NH_3Cl^+ is an acid-catalyzed reaction (Trofe et al. 1980), it is very unlikely to form NH_2Br and NHBrCl through the NH_3Cl^+ pathway. Standard Methods (4500-F) states that "...bromine, bromamine, and iodide react with DPD indicator and appear with free chlorine...." No apparent color change was observed in the samples during DPD titrations in the absence of KI catalyst, indicating that there was no NH_2Br or other bromamines detectable with the DPD technique used in this study. Even if NH_2Br had formed at very small amounts, it did not have any impact on NH_2Cl decay and HAA formation since no brominated HAA specie was detected during the experiments. These findings suggest that HAA formation at pH 9 is mainly a result of direct reactions between NH_2Cl with DOM.

DCAA was the only HAA specie detected at pH 9. No brominated HAA specie was detected during the course of experiments. It should be noted that the molecular weight of DCAA is 129 g/mol, thus a 8 nM difference in molar yield corresponds to only about 1 μg/L on a weight basis, an error within the variability of the analytical measurements. The 72-hr DCAA yields of

Table 5.2
72-hr normalized NH_2Cl consumption (i.e., $NH_2Cl_{consumed}/NH_2Cl_{initial}$) during chloramination experiments

Target Br/DOC (μg/mg)	pH 6			pH 7.5			pH 9		
	HPO	TPH	HPI	HPO	TPH	HPI	HPO	TPH	HPI
Greenville water									
50	0.77	0.84	0.65	0.47	0.31	0.32	0.27	0.27	0.31
130	0.80	0.82	0.71	0.47	0.38	0.32	0.27	0.26	0.31
350	0.90	0.95	0.89	0.50	0.42	0.37	0.27	0.27	0.30
Myrtle Beach water									
50	0.80	0.91	0.86	0.53	0.37	0.25	0.27	0.31	0.17
130	0.87	0.84	0.78	NA	NA	NA	NA	NA	NA
350	0.93	0.87	0.94	0.53	0.44	0.30	NA	NA	NA

NA: Not available since experiments were not performed at this condition.

Table 5.3
72-hr normalized $NHCl_2$ formation ($NHCl_{2,formed}/NH_2Cl_{initial}$ both as Cl_2) during chloramination at pH 6

Target Br/DOC (μg/mg)	Greenville water			Myrtle Beach water		
	HPO	TPH	HPI	HPO	TPH	HPI
50	0.2	0. 3	0.1	0.2	0.2	0.2
130	0.1	0.1	<MRL	0.1	0.1	0.1
350	0.1	<MRL	<MRL	0.1	0.1	<MRL

MRL: minimum reporting level.

the fractions were in the order of HPO>TPH ≈ HPI for both DOM sources (Table 5.4 and Table 5.5). Moreover, DCAA formation with the HPO fraction occurred faster, usually after 1 to 2 hr, than formation with TPH and HPI fraction, which occurred at about 4 to 8 hours (Figure 5.7 and Figure 5.8). These observations suggest that the HPO fractions in these waters are more reactive toward HAA formation than the TPH and HPI fractions.

pH 7.5 Results

A higher degree of NH_2Cl decay for all fractions was observed at pH 7.5 as compared to the results for pH 9. NH_2Cl decay and 72-hr NH_2Cl consumption were greater for the HPO fraction than the TPH and HPI fractions (Figure 5.9, Figure 5.10, Table 5.2). The TPH and HPI fractions exhibited relatively similar NH_2Cl decay patterns. The presence of bromide did not have any significant impact on the NH_2Cl decay in all fractions. There was no detectable amount of $NHCl_2$ measured during the experiments.

Table 5.4
72-hr HAA yields of Greenville DOM fractions

Target Br/DOC (μg/mg)	HAA	pH 6			pH 7.5			pH 9		
		HPO (nM)	TPH (nM)	HPI (nM)	HPO (nM)	TPH (nM)	HPI (nM)	HPO (nM)	TPH (nM)	HPI (nM)
50	DXAA	204	103	135	97	63	63	53	34	34
	TXAA	<MRL	<MRL	<MRL	<MRL	<MRL	<MRL	<MRL	<MRL	<MRL
	HAA_9	214	103	135	97	63	63	53	34	34
	n	0.34	0.52	0.89	0	0	0.15	0	0	0
130	DXAA	329	183	170	114	67	72	54	30	39
	TXAA	<MRL	12	15	<MRL	<MRL	<MRL	<MRL	<MRL	<MRL
	HAA_9	329	195	185	114	67	72	54	30	39
	n	0.94	1.38	1.47	0.20	0.40	0.54	0	0	0
350	DXAA	446	248	223	115	86	97	54	33	33
	TXAA	18	35	14	<MRL	<MRL	<MRL	<MRL	<MRL	<MRL
	HAA_9	501	306	254	115	86	97	54	33	33
	n	1.29	1.56	1.64	0.48	0.83	1.02	0	0	0

MRL: Minimum reporting level.
NA: Not available.
n: Bromine incorporation factor.

Table 5.5
72-hr HAA yields of Myrtle Beach DOM fractions

Target Br/DOC (μg/mg)	HAA	pH 6			pH 7.5			pH 9		
		HPO (nM)	TPH (nM)	HPI (nM)	HPO (nM)	TPH (nM)	HPI (nM)	HPO (nM)	TPH (nM)	HPI (nM)
50	DXAA	257	159	210	118	80	84	64	37	47
	TXAA	<MRL	<MRL	<MRL	<MRL	<MRL	<MRL	<MRL	<MRL	<MRL
	HAA_9	299	202	210	128	88	84	64	37	47
	n	0.25	0.34	0.76	0	0	0.18	0	0	0
130	DXAA	360	266	241	NA	NA	NA	NA	NA	NA
	TXAA	<MRL	<MRL	8	NA	NA	NA	NA	NA	NA
	HAA_9	360	266	264	NA	NA	NA	NA	NA	NA
	n	0.81	1.01	1.24	NA	NA	NA	NA	NA	NA
350	DXAA	500	350	387	136	111	131	NA	NA	NA
	TXAA	21	23	49	<MRL	<MRL	<MRL	NA	NA	NA
	HAA_9	562	408	462	136	111	131	NA	NA	NA
	n	1.22	1.36	1.59	0.36	0.62	1.03	NA	NA	NA

MRL: Minimum reporting level.
NA: Not available.
n: Bromine incorporation factor.

DXAA was the only HAA specie formed by all fractions of both waters. Its formation was, in general, higher in the HPO fractions than the TPH and HPI fractions for both DOM at each bromide level (Table 5.4, Table 5.5, Figure 5.9, and Figure 5.10). The fractions of MB water had higher DXAA yields as compared to the corresponding fractions of the GV water. Before 24-hr reaction time, similar amounts of DXAA were formed for each fraction of both waters independent of bromide level. After 24 hours, the HAA formation profiles of different bromide levels started to diverge and brominated species became more noticeable (Figure 5.9 and Figure 5.10). This pattern was more pronounced for the HPO fractions than for the THP and HPI fractions due to the higher reactivity of the HPO fraction. The slow kinetics of formation of brominated DXAA (i.e., a lag period of about 8 hr or higher for the appearance of BCAA and DBAA even at the highest Br/DOC ratio) is attributed to the relatively slow reaction of NH_3Cl^+ and HOCl formation at pH 7.5 (Figure 5.11). The amount of NH_3Cl^+ and HOCl formed was too small at the early reaction times to form enough NHBrCl that could be involved in the formation of brominated HAA. The calculated bromine incorporation factors showed that the affinity for bromine incorporation was in the order of HPI>TPH>HPO (Figure 5.11). This was due to the more effective incorporation of bromide by DOM with a more hydrophilic character. When the bromine incorporation of the same type fractions for MB and GV water were compared, the HPI fractions showed relatively similar incorporation, whereas bromine incorporation by the HPO and TPH fractions of GV water were higher as compared to those of the MB water.

Hwang et al. (2000) chloraminated various fractions of Colorado River water, a very low-$SUVA_{254}$ water (1.5 L/mg-m) under UFC condition (pH 8, residual chloramine 2 mg/L after 24-hr incubation time), which was very similar to this study. They observed that the HPI fraction had significantly higher HAA yields than the HPO and TPH fractions. The HAA yield of the Colorado River HPI fraction (0.0004 μmol/μmol = 33 nM/mg DOC) was much higher than the yields of GVTPH and GVHPI (4.5 nM/mg DOC) fractions, despite all three had similar $SUVA_{254}$ values. In this project, the MB HPO fractions showed the highest DXAA yield among all the fractions (10 nM/mg DOC), but still significantly lower as compared to the HPO fraction of Colorado River. These comparisons suggest that the DOM components in the very low $SUVA_{254}$ Colorado River water had significantly higher reactivity toward HAA formation than the waters tested in this study.

pH 6 Results

A significantly higher degree of NH_2Cl decay was observed at pH 6 than pH 7.5 and 9 with measurable amounts of $NHCl_2$ formed for all fractions (Figure 5.12, Figure 5.13, Table 5.3). The HPO fraction exhibited higher NH_2Cl consumption and resulted in more $NHCl_2$ than the HPI and TPH fractions. The NH_2Cl decay and the amount of NH_2Cl consumption increased with increasing bromide concentration, while the formation of $NHCl_2$ decreased with increasing bromide. This was probably because HOCl reacted with bromide to produce HOBr, resulting in less HOCl available for $NHCl_2$ formation.

All reactions were faster at pH 6 than pH 7.5 and 9. The fast NH_2Cl decay also resulted in significantly higher and faster DXAA formation compared to the other two pH values tested. The HPO fractions showed the highest DXAA yield for both DOMs, and the DXAA yields were in the order of HPO>TPH ≈ HPI (Figure 5.12, Figure 5.13, Table 5.4, Table 5.5). The comparison of DXAA formation for these two waters, including all fractions and RO isolate, is shown in Table 5.6. It is clear that the DOM components in MB water were more reactive than GV water. It is also evident that the reactivity of the water (i.e., RO isolates) was mainly controlled by the HPO

Table 5.6
72-hr DXAA molar yield comparison of Greenville and Myrtle Beach DOM fractions and RO isolates

Target Br/DOC	$DXAA_{MB}$ / $DXAA_{GV}$			
(μg/mg)	RO	HPO	TPH	HPI
50	1.20	1.26	1.54	1.56
130	1.25	1.09	1.45	1.42
350	1.14	1.12	1.41	1.74

fractions. This can be explained by the fact that the dominant fraction in both DOMs was the HPO fraction, which accounted for about 60–70% of the total DOM.

An increase in the bromide concentration accelerated the formation of HAA and shifted its speciation. Although DXAA was still the dominant HAA species, formation of small amounts of brominated TXAA (without any TCAA) was also observed with increasing bromide concentrations for all fractions (Table 5.4 and Table 5.5). The bromine incorporation was also very fast especially with increasing bromide concentration (Figure 5.14). The formation of brominated species was observed within 1 hr of reaction time, suggesting the strong substitution capacity of brominated oxidants. In a similar trend to pH 7.5 but with a higher magnitude and faster kinetics, bromine incorporation was observed in the order of HPI ≈ TPH>HPO for the GV water, while it was HPI >TPH>HPO for the MB water (Figure 5.14).

Bromine incorporation of the GV fractions was greater than the corresponding fractions of the MB water, a result consistent with the more hydrophilic character of the GV fractions. As discussed in the literature review, this is because brominated oxidants are better substitution agents than chlorinated ones, although their concentrations are usually much lower than the latter. In high $SUVA_{254}$ waters where the reactive DOM sites (aromatic moieties of DOM) are not limited, chlorinated oxidants can compete with brominated ones for DOM sites since their concentrations are much higher than the latter. However, in a low $SUVA_{254}$ water, the number of reactive DOM sites is limited. As a stronger substitution agent, brominated oxidants can be incorporated into less reactive sites (e.g., some aliphatic moieties of DOM) where chlorinated oxidants can not react. As a result, waters with high $SUVA_{254}$ show a lower bromide incorporation factor than low $SUVA_{254}$ waters. This is consistent with the observations of Liang and Singer (2003) during chlorination, however, the explanation provided here is somewhat different from their interpretation. They fractionated five surface waters into the HPO and HPI fractions using only XAD-8 resin. The chlorination results of these fractions showed that the HPI fractions exhibited higher bromide incorporation than the HPO fractions. Usually the HPO fractions contain a larger amount of activated aromatic moieties as compared to the HPI fraction, which is rich in aliphatic structures, Liang and Singer (2003) concluded that HOBr was more reactive with aliphatic precursors than with aromatic precursors while HOCl was more reactive with the aromatic components of DOM. Although the BIF was higher in the HPI fractions, it only means the ratio of brominated to chlorinated HAA is higher. However, the absolute formation of HAA, including both brominated and chlorinated species, was less in the HPI fraction than in the HPO fraction. This observation is similar to the results in this study during chloramination. These results indicate that both kinds of oxidants, HOCl and HOBr in chlorination or NH_2Cl and NHBrCl in chloramination, are more reactive with aromatic moieties of DOM than with aliphatic moieties. As better oxidants and substitution agents, brominated oxidants (HOBr and NHBrCl) can attack more DOM sites besides aromatic sites, so the formation of brominated HAA is less sensitive to the aromatic content concentration of DOM (i.e., $SUVA_{254}$) than the formation of chlorinated HAA.

Although bromide increased both NH_2Cl decay and DXAA formation, the bromide effect was more significant in NH_2Cl decay than in DXAA formation (Figure 5.15). The profiles of NH_2Cl consumption of three fractions at each bromide level were relatively comparable, especially for GV water at high bromide levels. However, DXAA formation by the HPO fractions showed much greater formation than the other two fractions. It appears that although the presence of bromide significantly increased NH_2Cl decay in hydrophilic fractions, major amounts of NH_2Cl were consumed without leading to DXAA formation due to the limited DOM reactive sites. These results indicated the DXAA formation in the HPI fractions was controlled by the limited reactive DOM sites as compared to HPO fractions.

At pH 6, the correlation between $SUVA_{254}$ of DOM fractions and RO isolates with HAA at ambient bromide levels after 24-hr chlorination and chloramination is shown in Figure 5.16. It is apparent that $SUVA_{254}$ was a good surrogate parameter to assess overall HAA formation. As expected, the impact of DOM characteristics was more significant in chlorination than chloramination. The difference in HAA formation between these two disinfection methods was more pronounced for high $SUVA_{254}$ DOMs, indicating chloramination is likely to provide more benefit for controlling HAA in waters with hydrophobic character (i.e., high SUVA) than hydrophilic character.

In the presence of bromide during chloramination, after 72-hr incubation time, the DXAA formation and the slope of DXAA formation curve increased as bromide concentration increased (Figure 5.17a), indicating the role of brominated oxidants in DXAA formation. The correlation was much better when HPI fractions were omitted since the $SUVA_{254}$ values of TPH and HPI were very close for MB water and HPI showed even higher $SUVA_{254}$ than TPH in GV water (Figure 5.17b).

Undoubtedly, an increased bromide concentration enhanced the bromine incorporation; however, this effect started to level off after reaching a certain level (Figure 5.18). The fractions with $SUVA_{254}$ values < 2 L/mg-m, showed higher BIF, about 0.3 units, than the fractions with $SUVA_{254}$ > 2 L/mg-m. The fractions with $SUVA_{254}$ values from 2 to 5 L/mg-m had similar BIF at each Br/DOC ratio. This indicated that once the $SUVA_{254}$ value was above a certain level, around 2 L/mg-m in this study, the active sites of DOM fractions were not limited as compared with those DOM fractions with low $SUVA_{254}$. Therefore, chlorinated oxidants can compete more effectively with brominated oxidants for DOM reactive sites forming DXAA as compared to low $SUVA_{254}$ DOM where only limited reactive sites were available. This also indicated that $SUVA_{254}$ may not be sensitive enough to be a good surrogate predictor for the bromide effect even though it is a good predictor of DOM characteristics and HAA reactivity.

Overall, for all bromide levels tested in this study, the order of 72-hr DXAA formation could be divided into three groups as follows: (MBHPO > GVHPO) > (MBTPH ≈ MBHPI) > (GVTPH ≈ GVHPI), again indicating that the hydrophobic component was the major precursor of DXAA formation (Figure 5.19). The effect of bromide for all fractions is shown in Figure 5.20. Undoubtedly, an increased bromide concentration enhanced the bromide incorporation, however, this effect started to level off after reaching a certain level (i.e., Br/DOC = 130 μg/mg). As expected, the order of BIF was the opposite of that for DXAA yield as follows: (GVHPI ≈ GVTPH) > MBHPI > (MBTPH ≈ GVHPO) > MBHPO. It was noticed that the profiles of TPH and HPI fractions of GV water were consistently very close to each other, indicating similar characteristics. This was because that GV water was a hydrophilic water; as a result, its TPH character was similar to its HPI unlike the MB fractions. Another possible reason was that the excessive amount of silica in the GV water adsorbed the relatively hydrophobic parts of the TPH fractions into HPO, resulting that the rest TPH captured by XAD-4 column was more hydrophilic than it was supposed to be.

SUMMARY OF IMPORTANT FINDINGS

Important findings from this phase of this study are as follows:

- The results obtained in this study and in previous investigations with preformed chloramines indicate that at least 80% of HAA to be formed from chloramines(s) in natural waters will be in the DXAA form, and the contribution of DXAA may reach and exceed over 90% in some cases. A small amount of MXAA formation may also be observed, while formation of TXAA is usually negligible.
- *pH effect:* DXAA formation both on a molar and mass concentration basis decreased with increasing pH at all three bromide concentrations in both waters. Formation of MXAA, although in very small amounts, appeared to decrease with increasing pH, while TXAA formation was usually negligible, except small amounts of TXAA formation was observed at pH 6 with the highest Br/DOC ratios. These trends were different from those observed during chlorination. For chlorination, formation of MXAA and DXAA is almost independent of pH at typical water treatment conditions (i.e., pH 6 to 9), while formation of TXAA decreases with increasing pH. The formation of DXAA was very sensitive to pH during chloramination. The pH effect was not proportional to the change in pH. There was larger decrease in DXAA yield when pH was increased from 6 to 7.5 than from 7.5 to 9.0. It is clear from these results that water utilities intending to decrease HAA formation during chloramination should maintain pH above 7.5.
- *Bromide effect:* Bromide effect was pH dependent. At pH 6, DXAA formation increased and shifted toward brominated species with increasing bromide concentration. Small amounts of MXAA formation were observed, which shifted also from MCAA to MBAA with increasing Br/DOC ratio. The presence of bromide resulted in TXAA formation at the highest Br/DOC ratios for both DOMs. At pH 7.5, the impact of bromide on DXAA formation and NH_2Cl decomposition was relatively small, and at pH 9, it was essentially negligible. Bromine incorporation was higher and occurred more rapidly at pH 6 than pH 7.5. No brominated specie was formed at pH 9. These results showed that one additional advantage of having an operational pH of 7.5 or above for chloramination is to reduce the formation of brominated HAA species.
- *DOM effect:* The aromatic content of DOM (i.e., $SUVA_{254}$) appeared to have some impact on the DXAA yields during chloramination at pH values (e.g., pH 6 tested in this study) that considerably increase NH_2Cl decomposition. However, the effect of $SUVA_{254}$ on the HAA yields is much less pronounced during chloramination than chlorination. In this study, at pH 6 the formation of DXAA in MB DOM ($SUVA_{254}$ = 4.9 mg/L-m) was about 10–30% higher (on a mass and mol basis) than in GV ($SUVA_{254}$ = 2.6 mg/L-m) and SP DOM ($SUVA_{254}$ = 3.2 mg/L-m) at three bromide levels. At pH 7.5 and 9.0, both GV and MB DOMs showed very similar DXAA yields. These results indicated that pH was a more important factor controlling the extent of

DXAA formation and speciation than $SUVA_{254}$ of DOM and bromide in typical drinking water conditions.

- *Reactivity of DOM fractions:* To investigate the effect of DOM characteristics, two surface waters with different $SUVA_{254}$ values were fractionated into HPO, TPH and HPI fractions. The results for the effects of DOM and bromide on NH_2Cl decay and DXAA formation are summarized in Table 5.7. Generally, with the decrease of pH, the effect of DOM characteristics and bromide became more important. At pH 9, the effect of DOM was minimal and the effect of bromide was not observed. At pH 7.5, both effects were observed but not to a significant degree. At pH 6, for all fractions and RO DOM, the DXAA yield increased with increasing $SUVA_{254}$. However, the effect of $SUVA_{254}$ on the HAA yields was less pronounced during chloramination than during chlorination, and the differences in DXAA yield between chlorination and chloramination increased with $SUVA_{254}$. The overall trend of DXAA formation was (MBHPO > GVHPO) > (MBTPH ≈ MBHPI)> (GVTPH ≈ GVHPI), indicating that the hydrophobic component was the major precursor of DXAA formation. For bromine incorporation, the opposite trend held true for the bromine incorporation factor (n): GVHPI ≈ GVTPH > MBHPI > MBTPH ≈ GVHPO > MBHPO. The bromine incorporation factor showed an inverse but not a linear correlation with $SUVA_{254}$, suggesting $SUVA_{254}$ alone may not be sensitive enough to predict the bromide effect even though it was a good indicator for HAA reactivity of DOM.
- The results obtained in this phase of the study indicate that utilities need to maintain the pH above 7.5 during operation to control DBP formation during chloramination, which is also helpful to maintain a relatively stable NH_2Cl residual in a distribution system. Although the HPI fractions were more reactive with bromide than the HPO fractions, the HPO fractions consistently showed higher DXAA yields than the relatively hydrophilic fractions. Thus, the enhanced removal of the hydrophobic component especially in moderate and high $SUVA_{254}$ waters can assist with better control of DBP formation during chloramination. Therefore, chloramination is expected to benefit from (enhanced) coagulation/flocculation which has been shown to preferentially remove hydrophobic DOM components.

Table 5.7
Effects of DOM and bromide on NH_2Cl decay and DXAA formation during chloramination

pH	NH_2Cl decay			DXAA formation	
	DOM effect	Bromide effect	$NHCl_2$	DOM effect	Bromide effect
pH 9	No, similar for all fractions	No, similar for all fractions	No formation	HPO > TPH ≈ HPI, HPO faster kinetics MB = GV for each fraction	None
pH 7.5	HPO > TPH ≈ HPI	Yes, not significant	No formation	HPO > TPH ≈ HPI MB > GV for each fraction	No brominated species until 8 hours later. No effect for TPH and HPO. HPI > TPH = HPO
pH 6	HPO > TPH ≈ HPI	Yes, significant	Yes, measurable amount	HPO > TPH ≈ HPI MB > GV for each fraction Overall, at pH 6, (MBHPO > GVHPO) > (MBTPH ≈ MBHPI) > (GVTPH ≈ GVHPI)	TXAA formed at high bromide level. Brominated species occurred within 1 hour. HPI ≈ TPH > HPO (GV) HPI > TPH > HPO (MB) GV > MB Overall, at pH 6, GVHPI ≈ GVTPH > MBHPI > MBTPH ≈ GVHPO > MBHPO

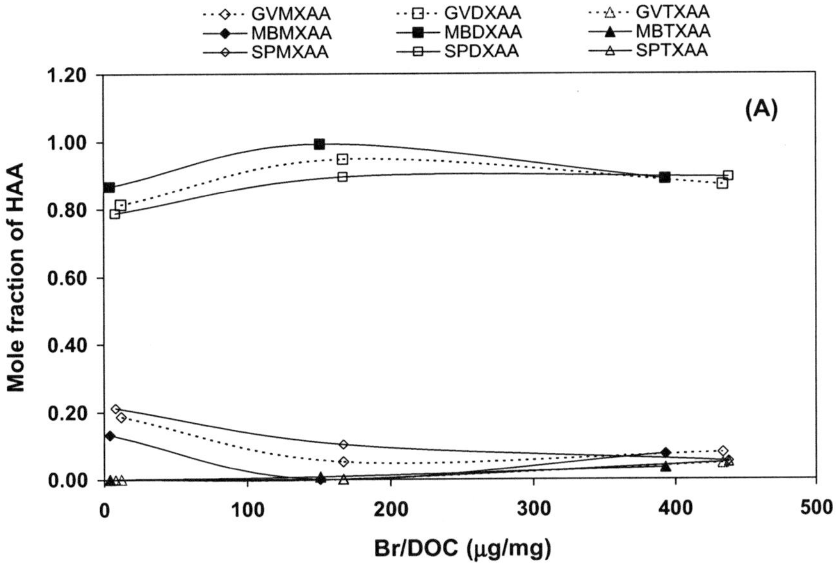

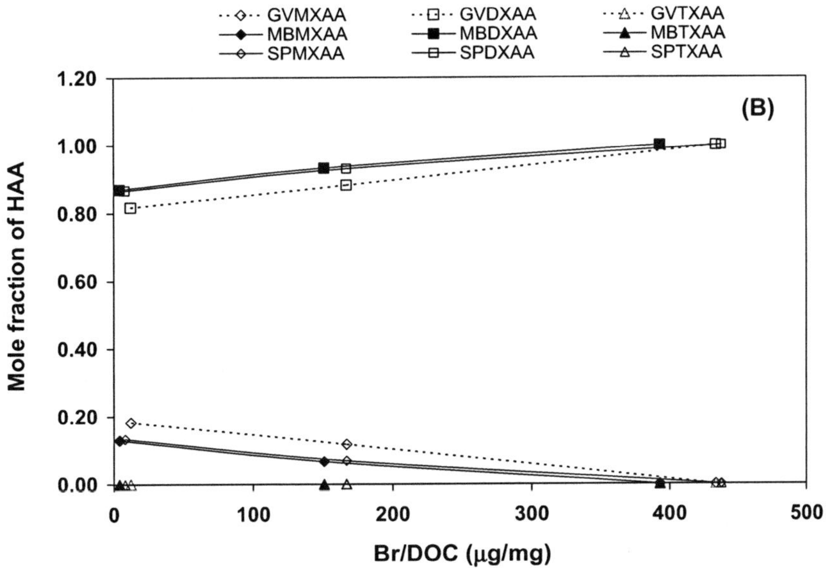

Figure 5.1 Distribution of MXAA, DXAA, TXAA at 72 hours during chloramination of Greenville and Myrtle Beach RO isolates at (A) pH 6 and (B) pH 7.5

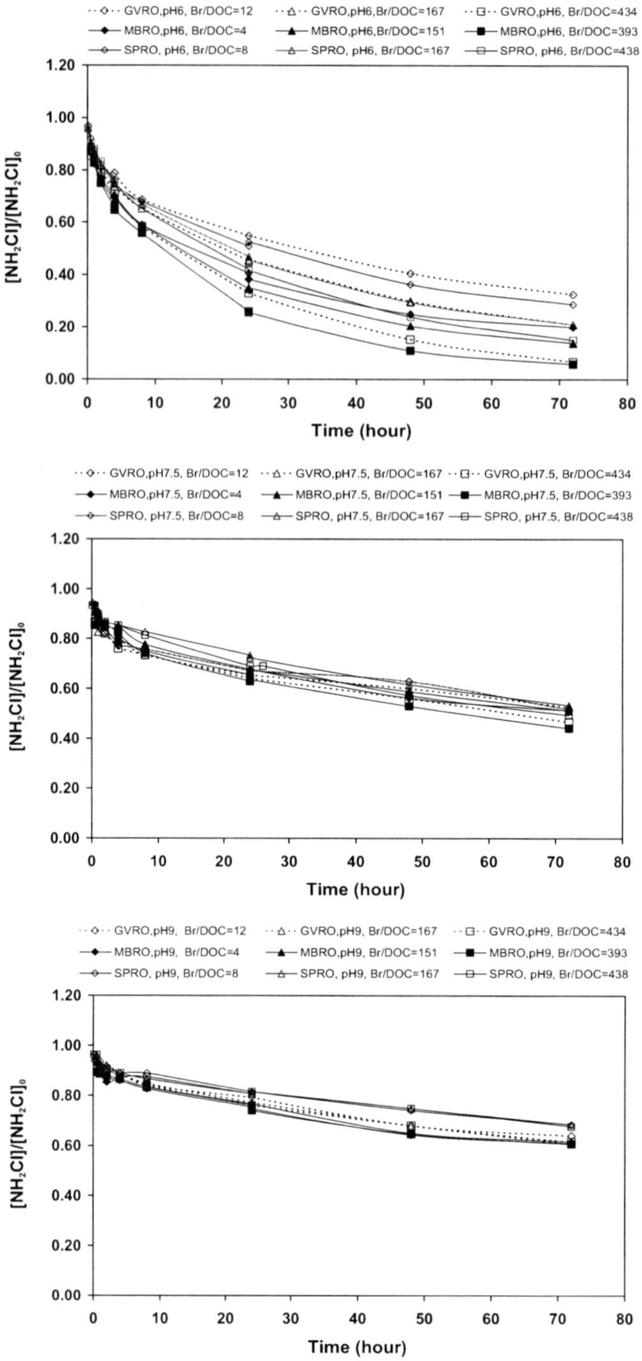

Figure 5.2 NH_2Cl consumption by Greenville, Spartanburg, and Myrtle Beach RO isolates

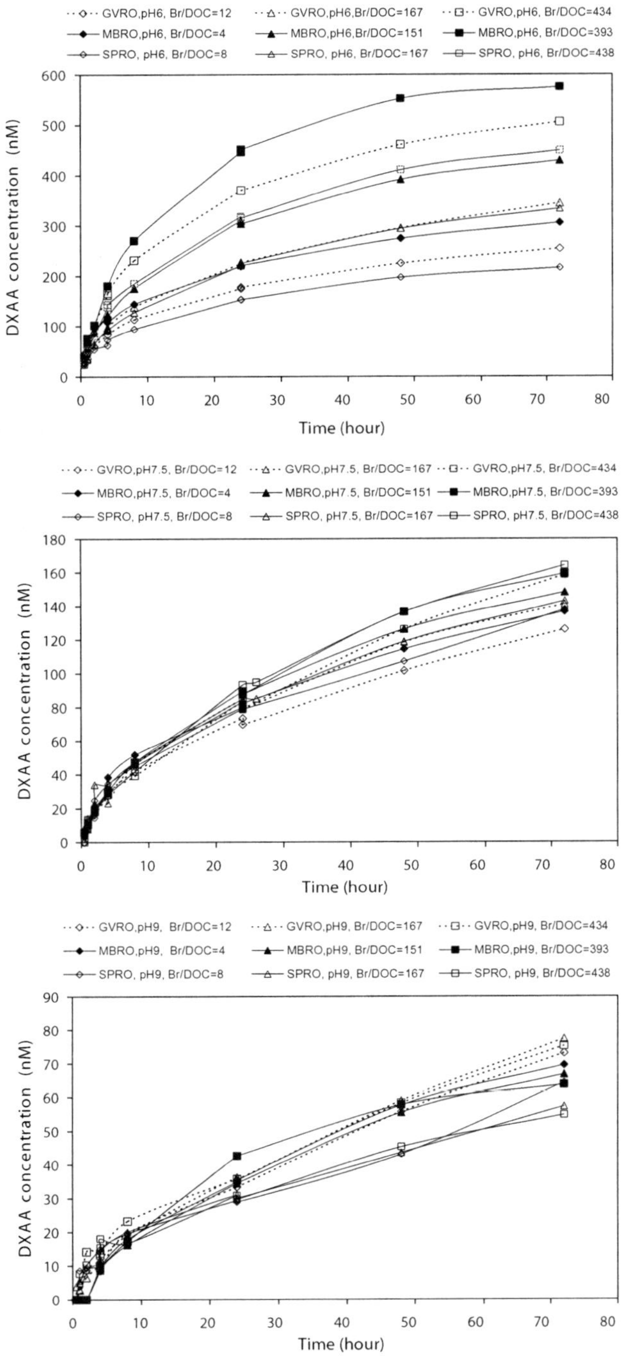

Figure 5.3 DXAA formation of Greenville, Spartanburg, and Myrtle Beach RO isolates

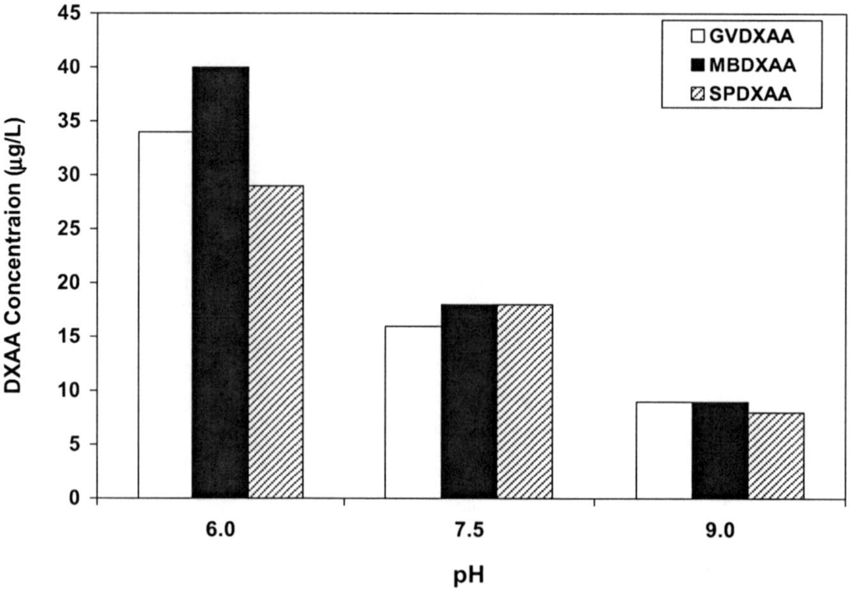

Figure 5.4 pH effect on 72-hr DXAA formation for Greenville, Spartanburg, and Myrtle Beach RO isolates at ambient Br/DOC (<12 µg/mg) level

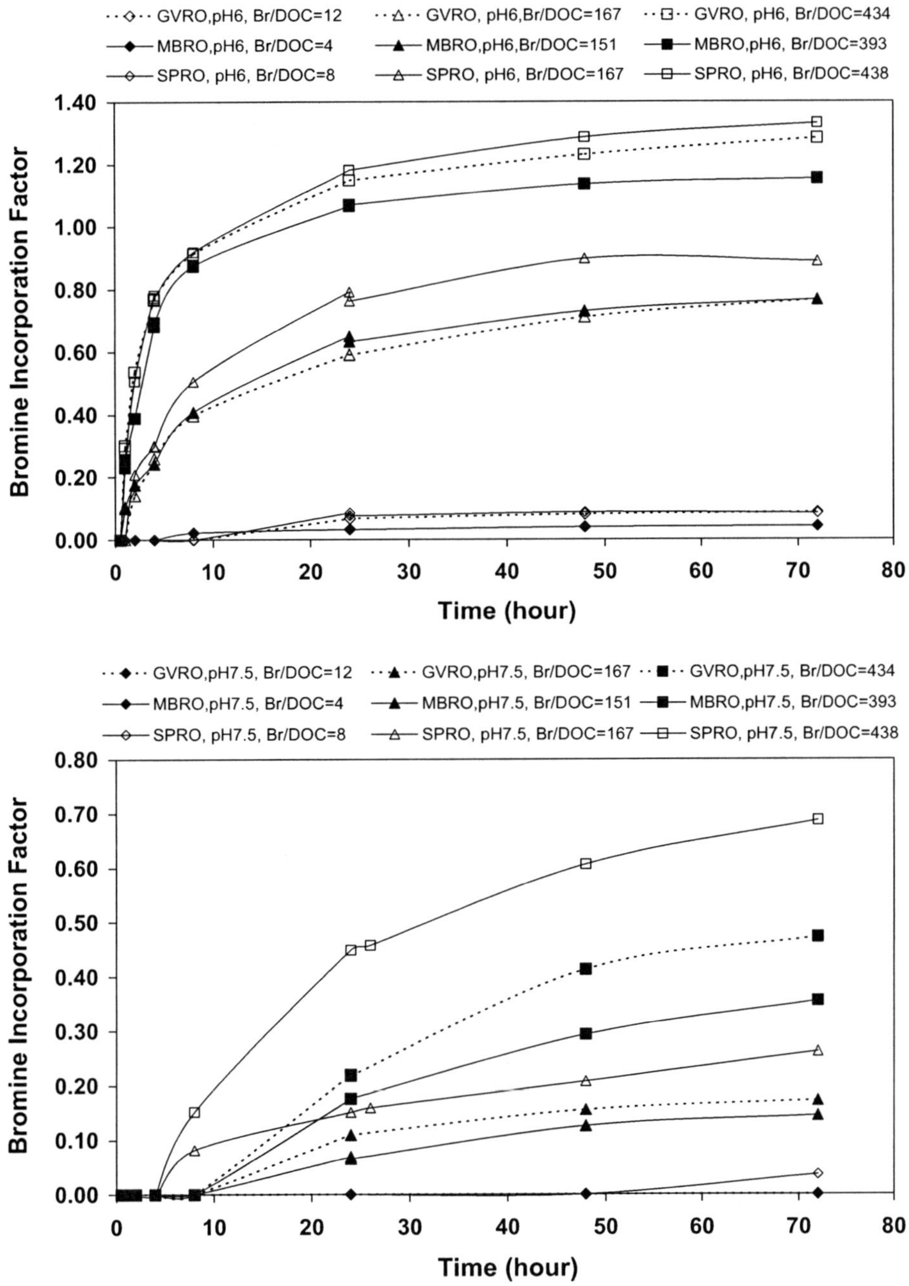

Figure 5.5 Bromine incorporation of Greenville, Spartanburg, and Myrtle Beach RO isolates

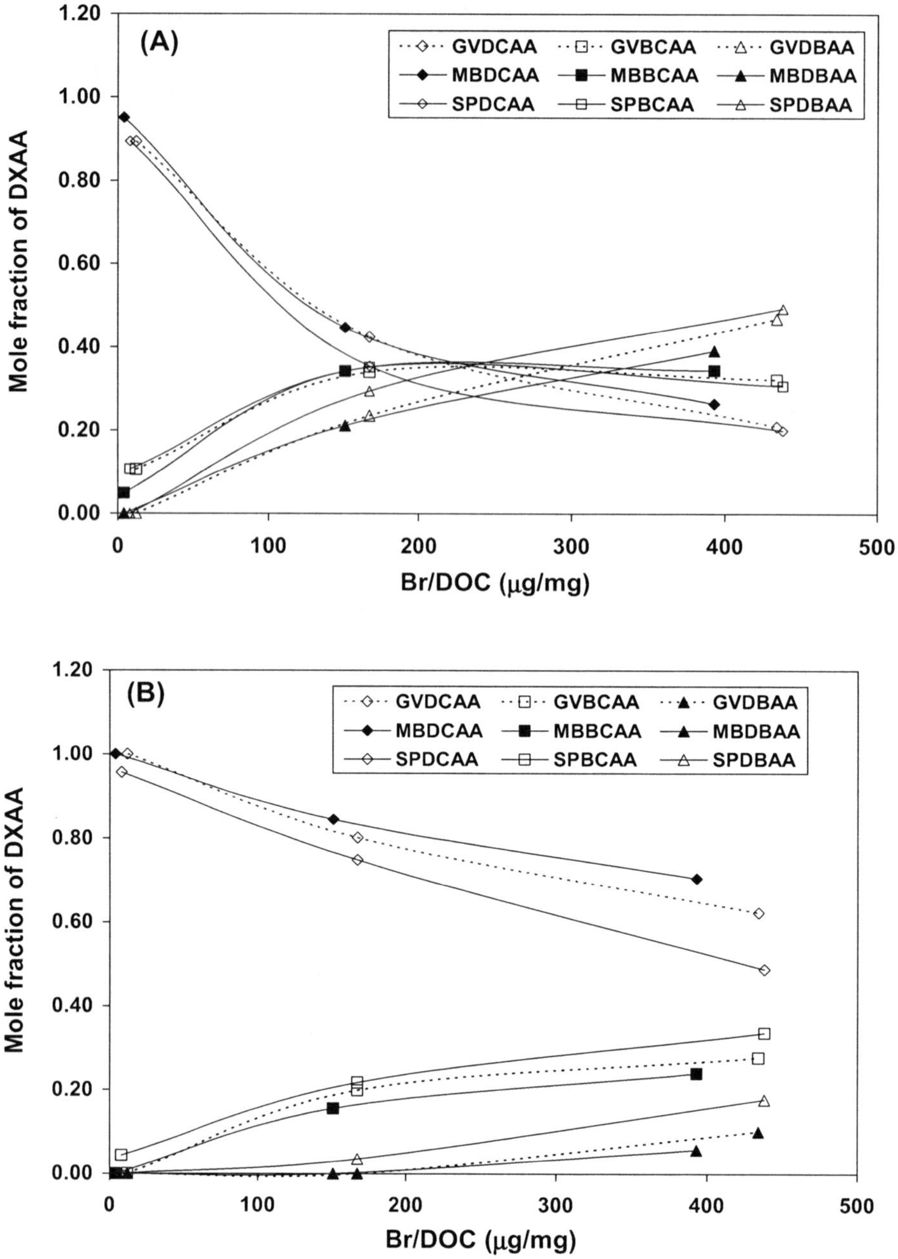

Figure 5.6 Distribution of individual DXAA species at 72-hr in Greenville, Spartanburg, and Myrtle Beach RO isolates at (A) pH 6 and (B) pH 7.5

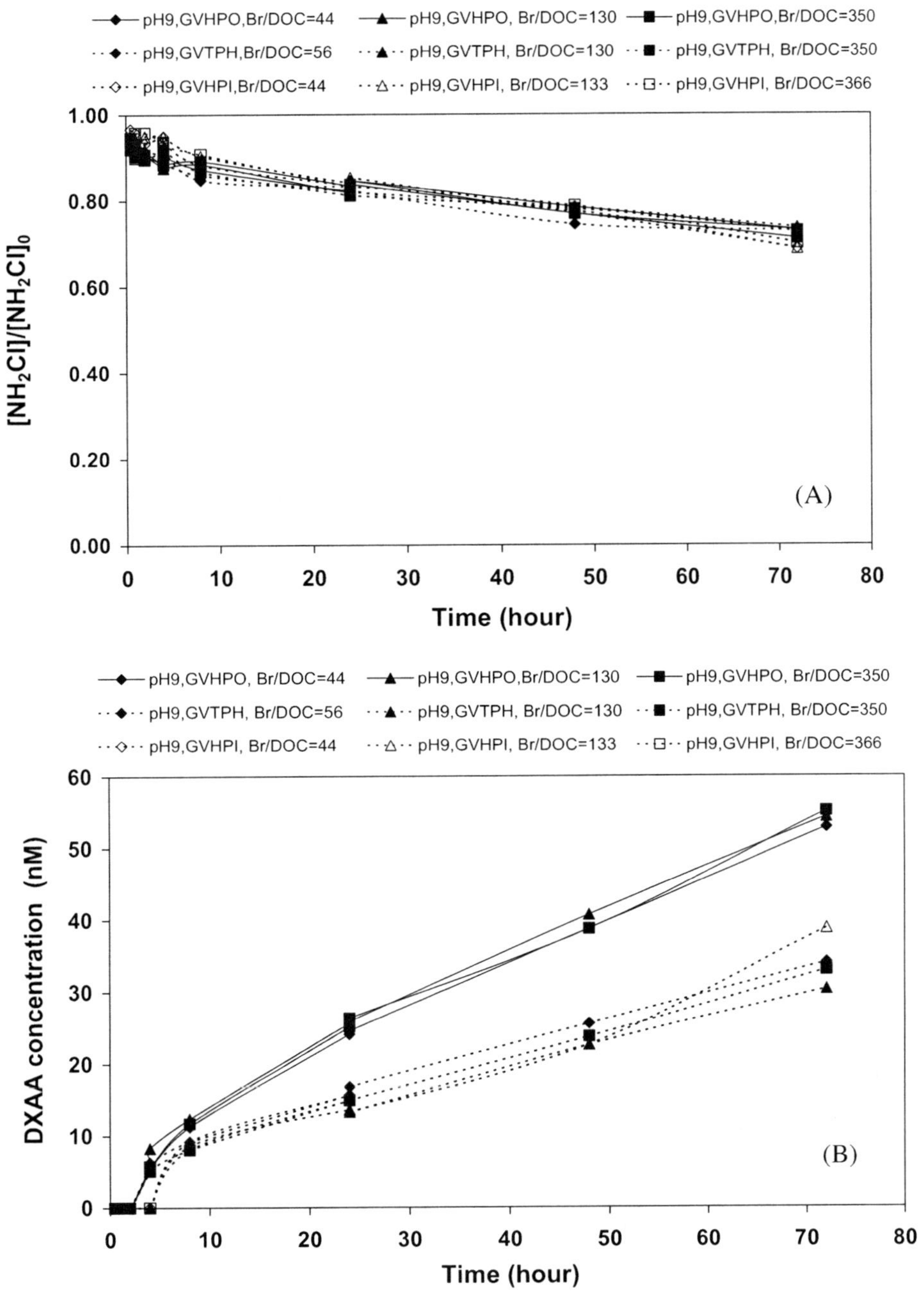

Figure 5.7 (A) NH_2Cl consumption and (B) DXAA formation during chloramination of Greenville DOM fractions at pH 9

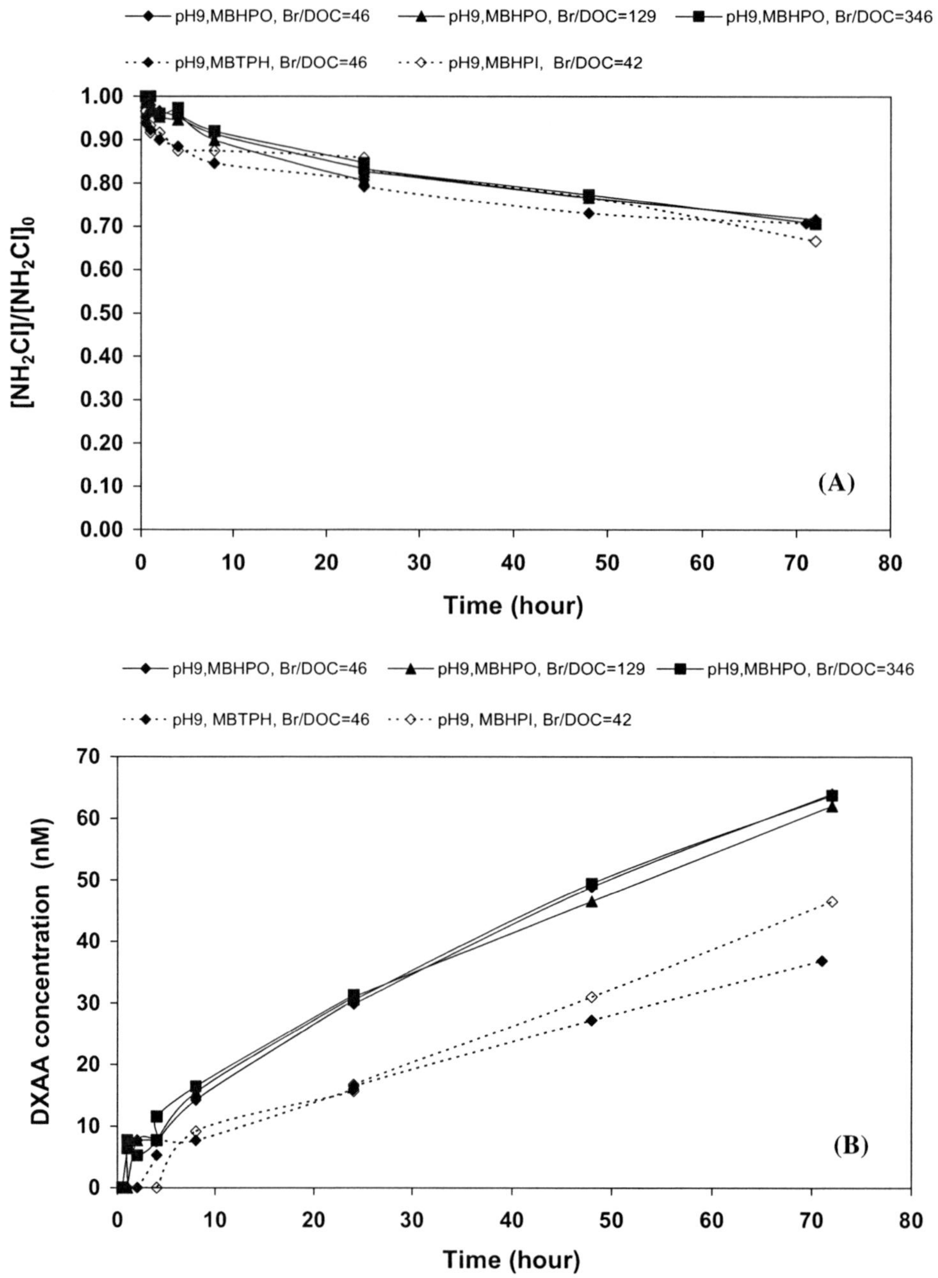

Figure 5.8 (A) NH_2Cl consumption and (B) DXAA formation during chloramination of Myrtle Beach DOM fractions at pH 9

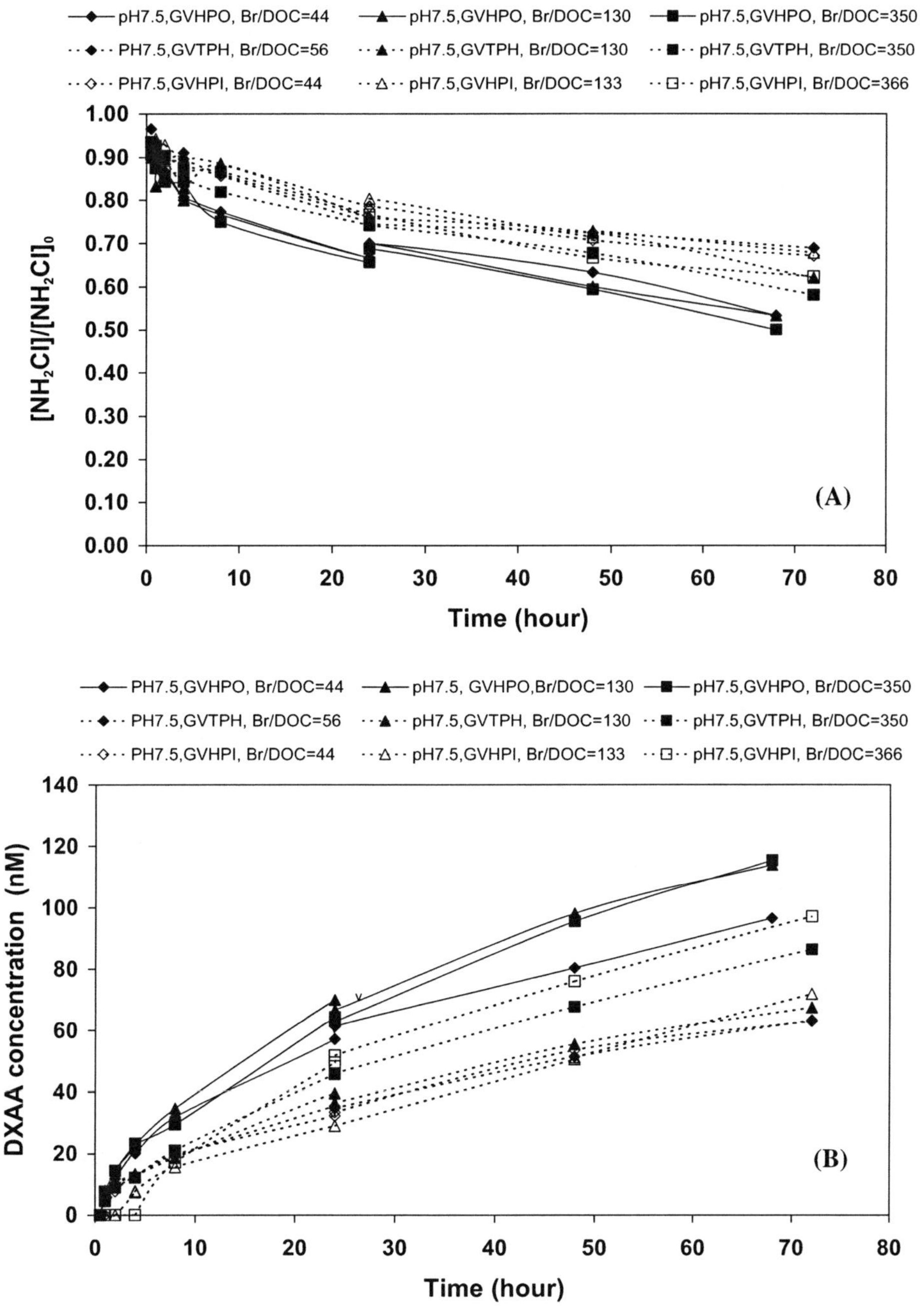

Figure 5.9 (A) NH_2Cl consumption and (B) DXAA formation during chloramination of Greenville DOM fractions at pH 7.5

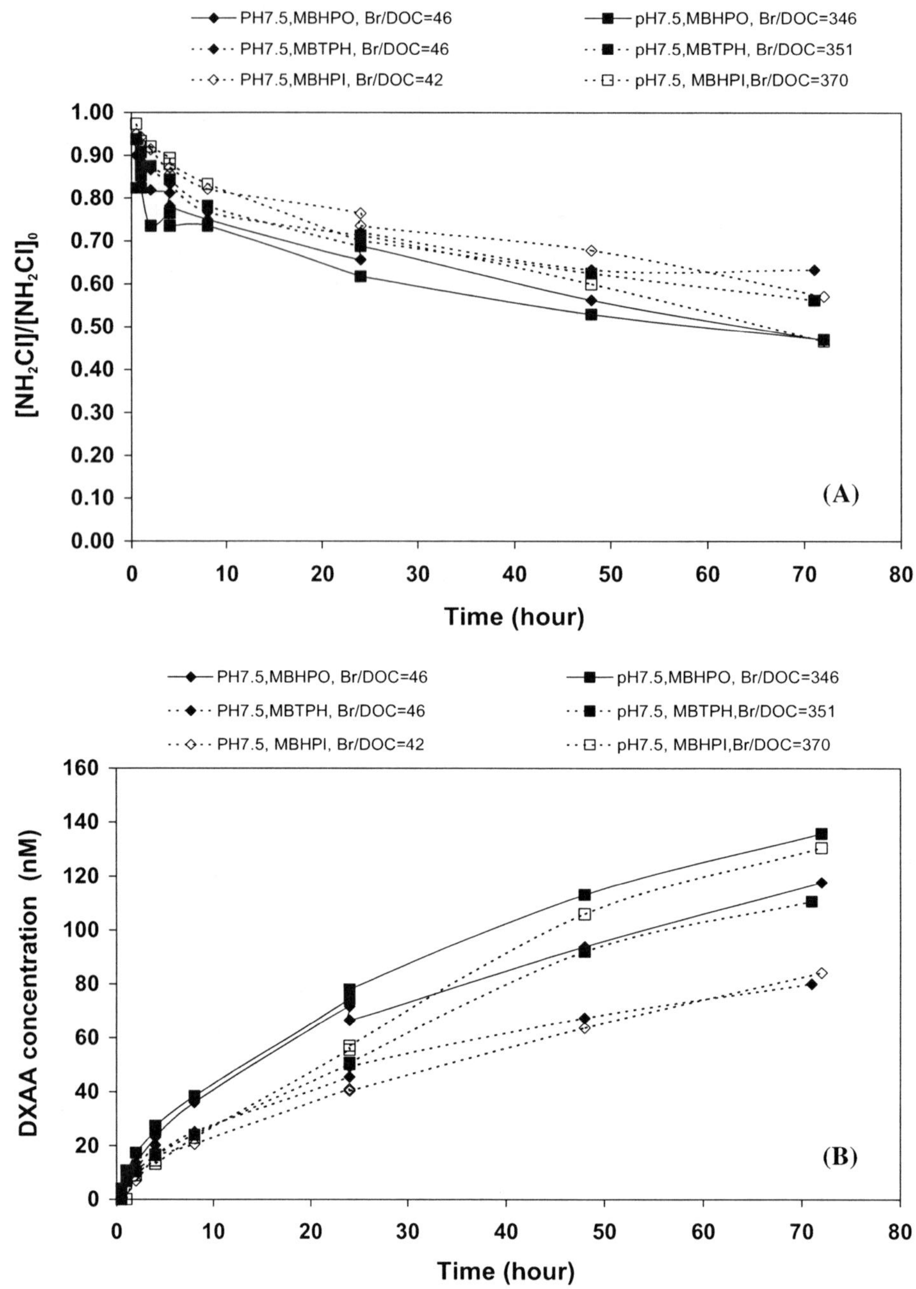

Figure 5.10 (A) NH_2Cl consumption and (B) DXAA formation during chloramination of Myrtle Beach DOM fractions at pH 7.5

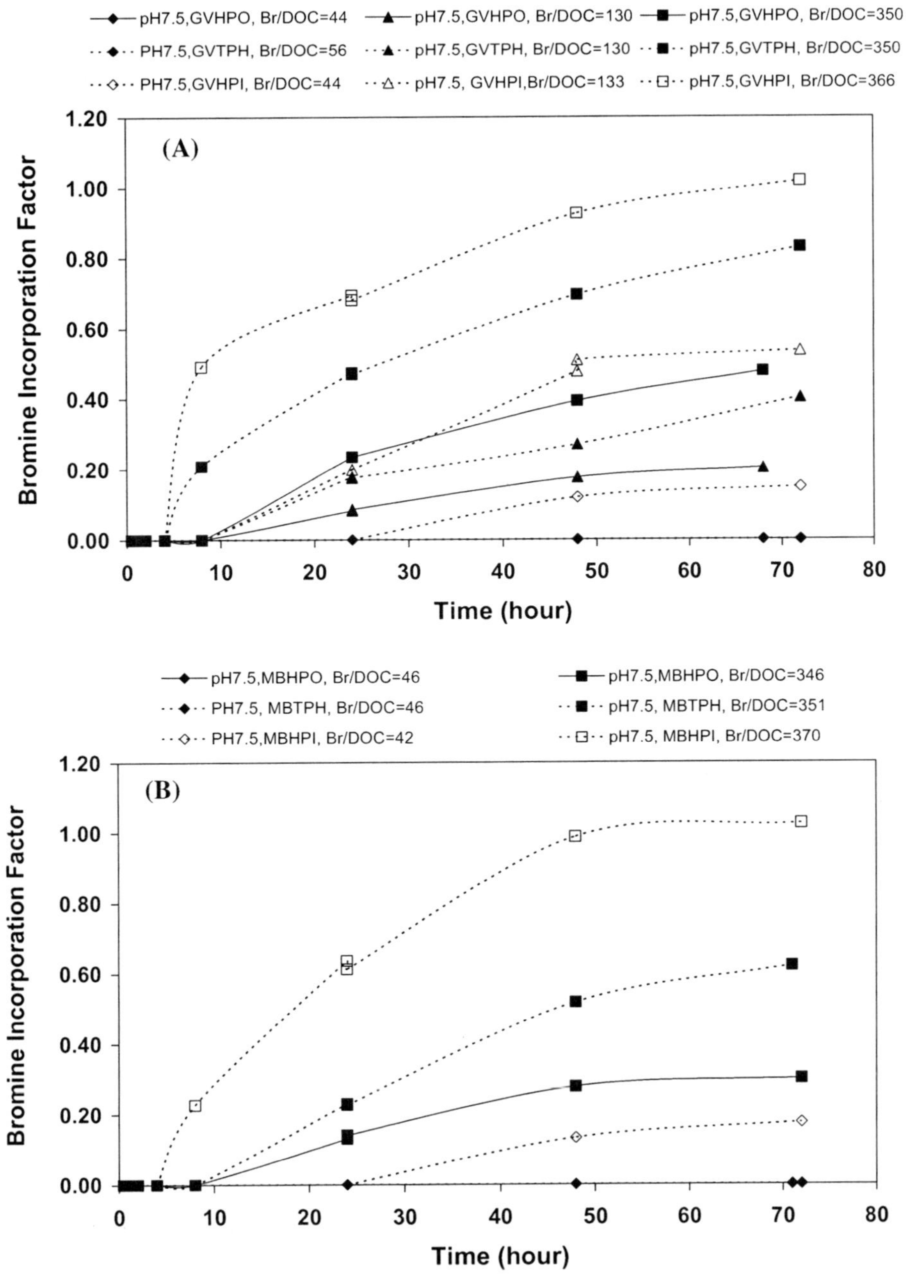

Figure 5.11 Bromine incorporation of (A) Greenville and (B) Myrtle Beach DOM fractions during chloramination at pH 7.5

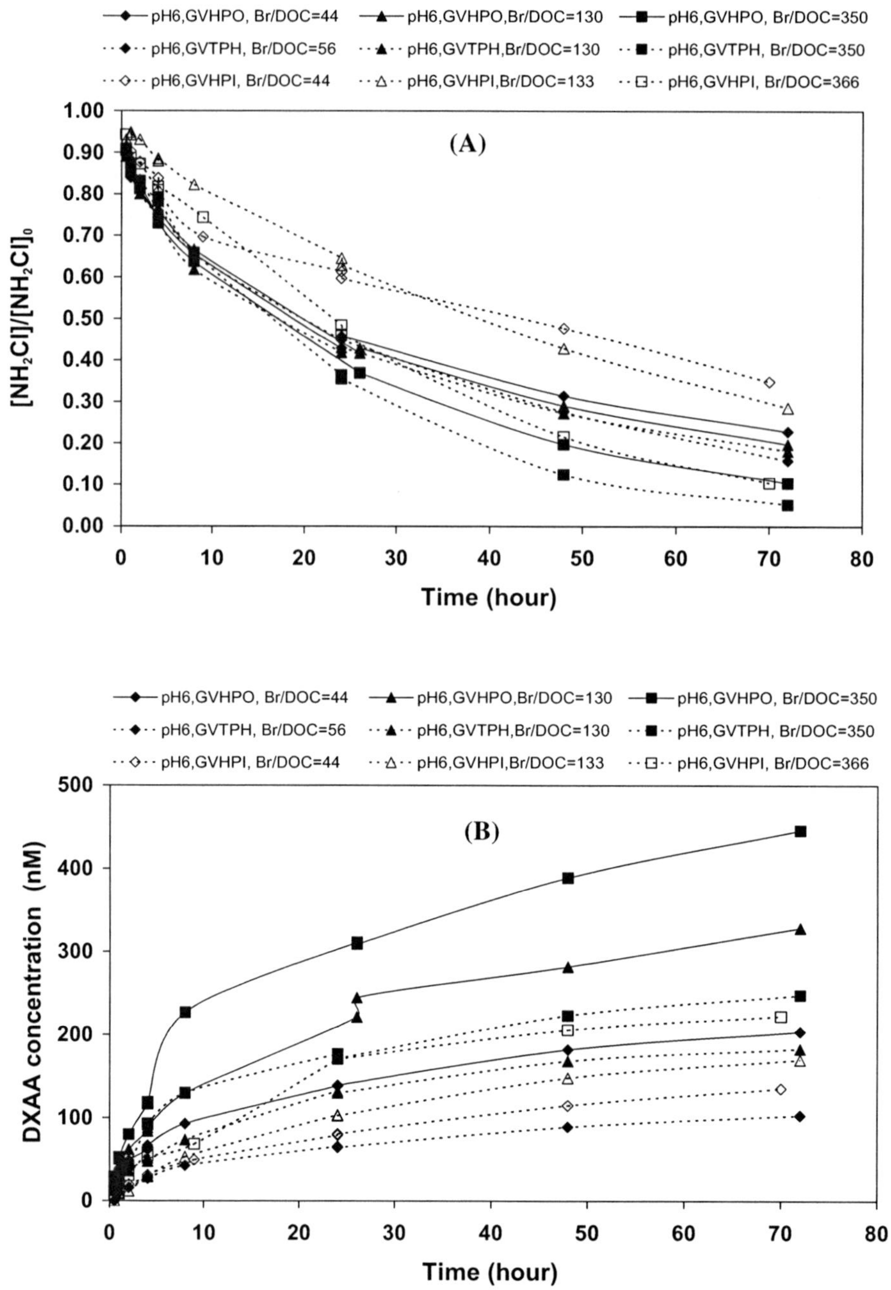

Figure 5.12 (A) NH_2Cl consumption and (B) DXAA formation during chloramination of Greenville DOM fractions at pH 6

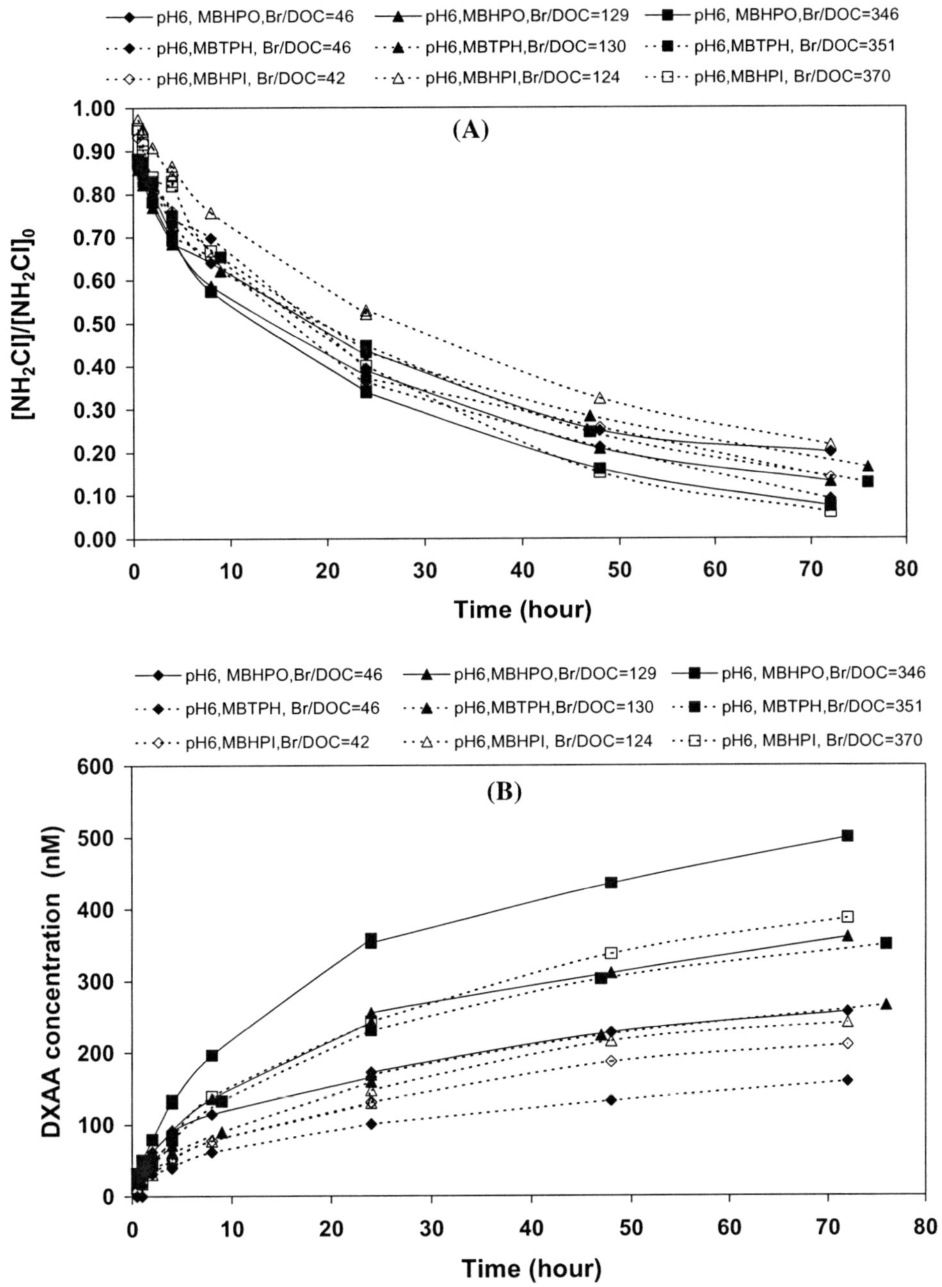

Figure 5.13 (A) NH_2Cl consumption and (B) DXAA formation during chloramination of Myrtle Beach DOM fractions at pH 6

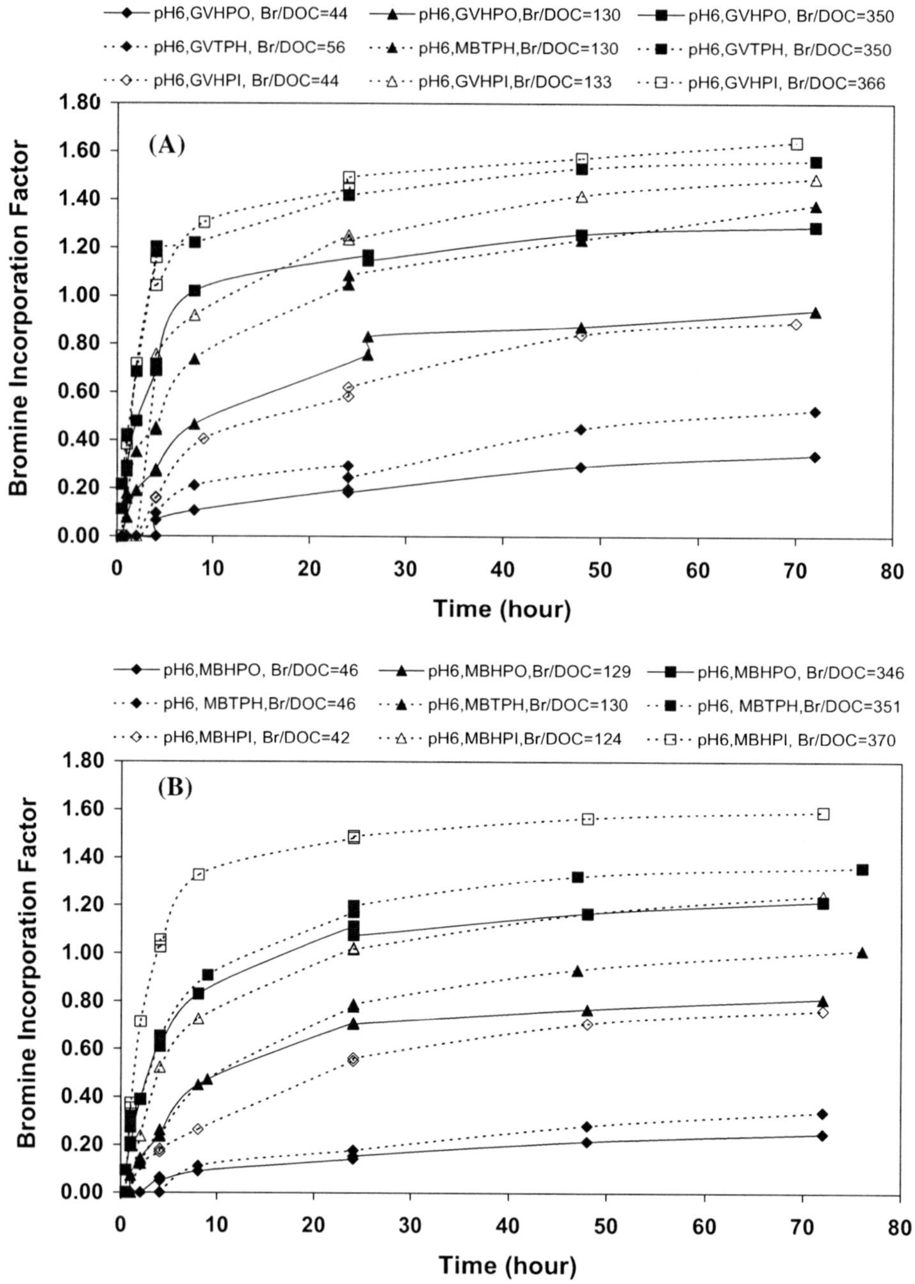

Figure 5.14 Bromine incorporation of (A) Greenville and (B) Myrtle Beach DOM fractions during chloramination at pH 6

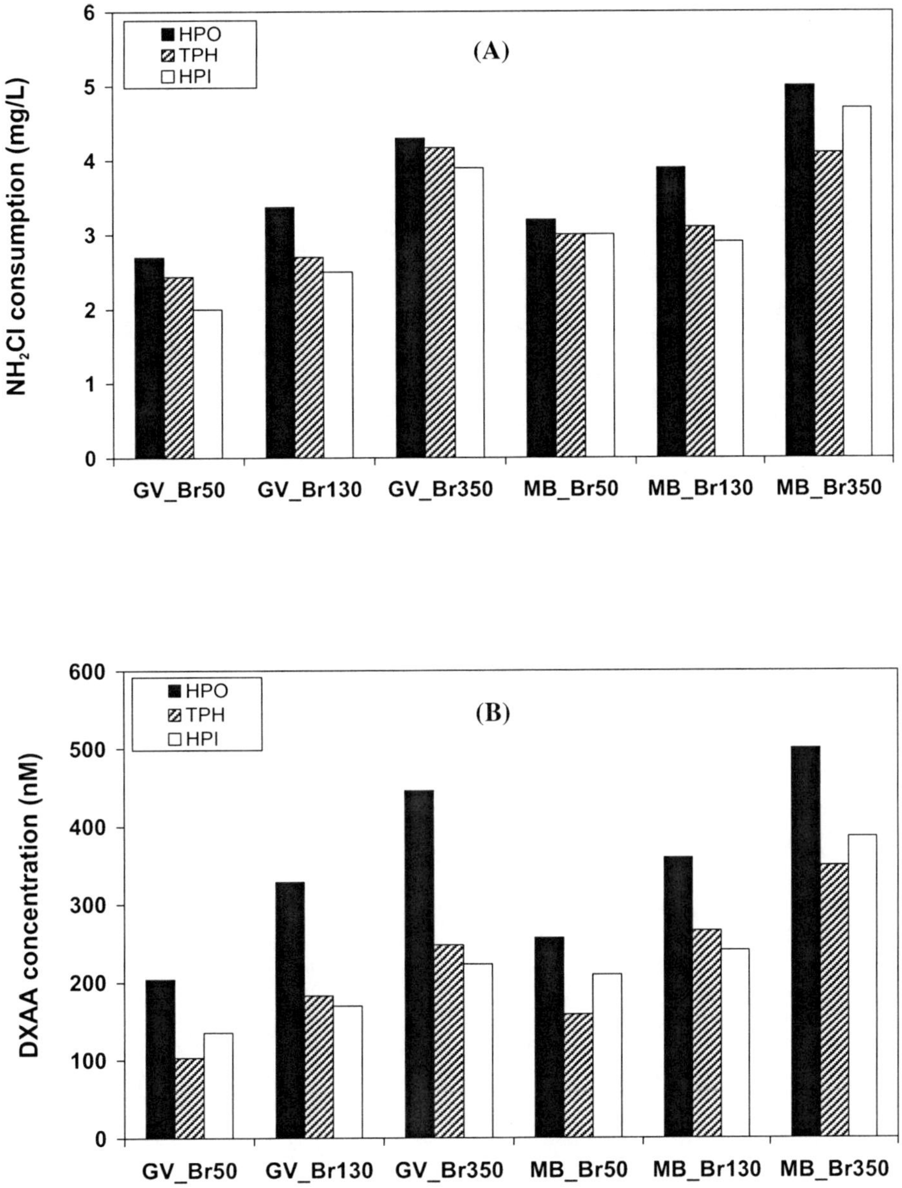

Figure 5.15 Comparison of 72-hr (A) NH_2Cl consumption and (B) DXAA formation of Greenville and Myrtle Beach DOM fractions at pH 6

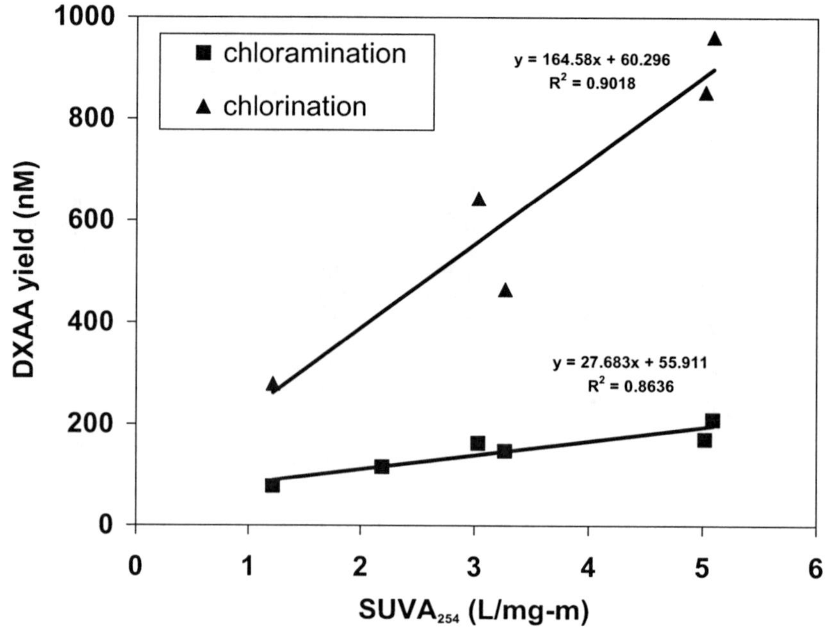

Figure 5.16 Effect of $SUVA_{254}$ on 24-hr DXAA yield for Greenville and Myrtle Beach RO isolates and DOM fractions at ambient Br/DOC levels during chlorination and chloramination at pH 6

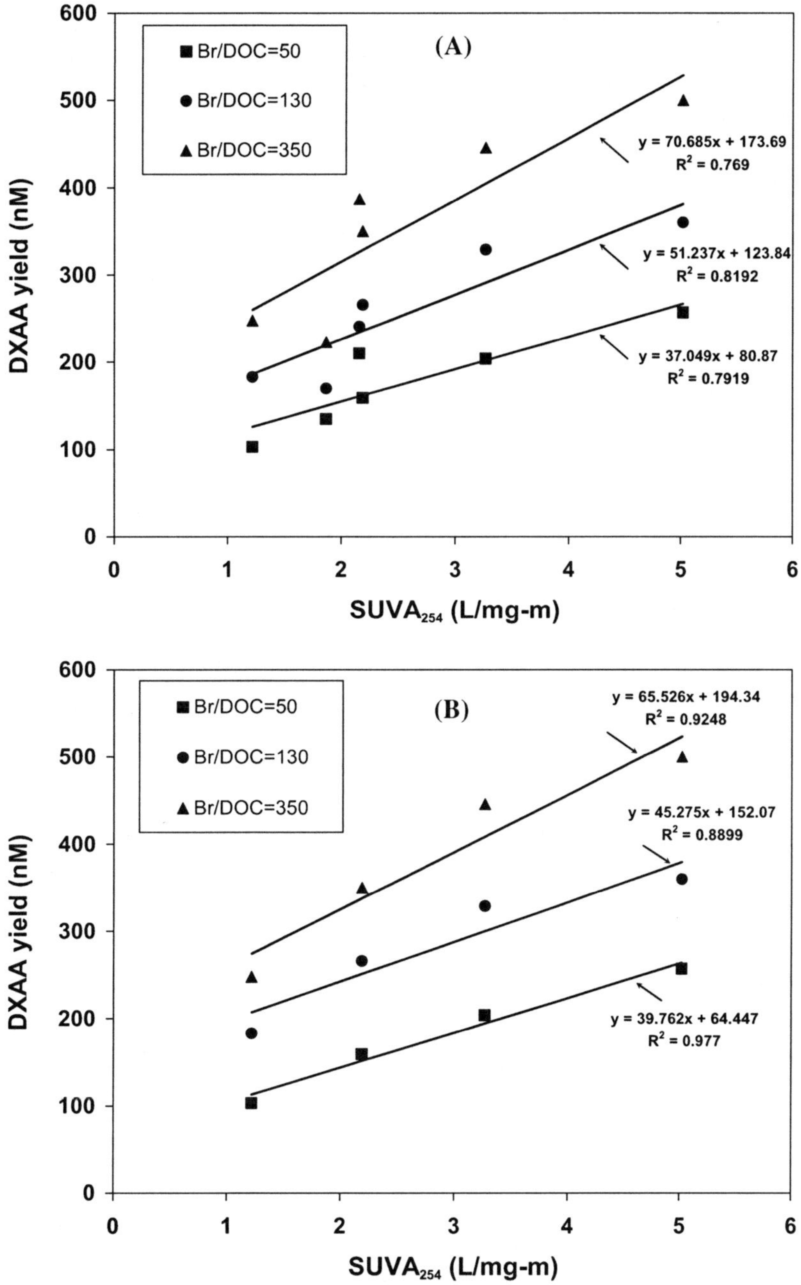

Figure 5.17 Effect of $SUVA_{254}$ on 24-hr DXAA yield for Greenville and Myrtle Beach RO isolates: (A) all DOM fractions and (B) all DOM fractions without HPI at pH 6

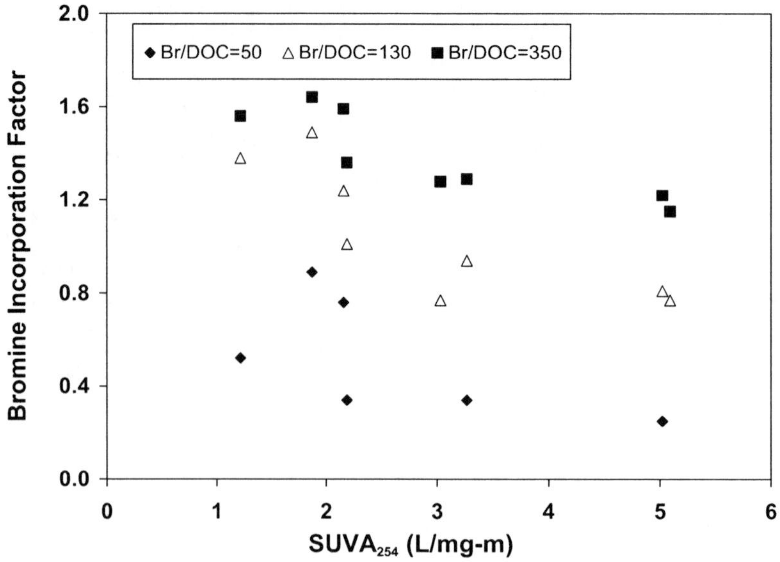

Figure 5.18 Bromine incorporation factors (72-hr) for all fractions at pH 6

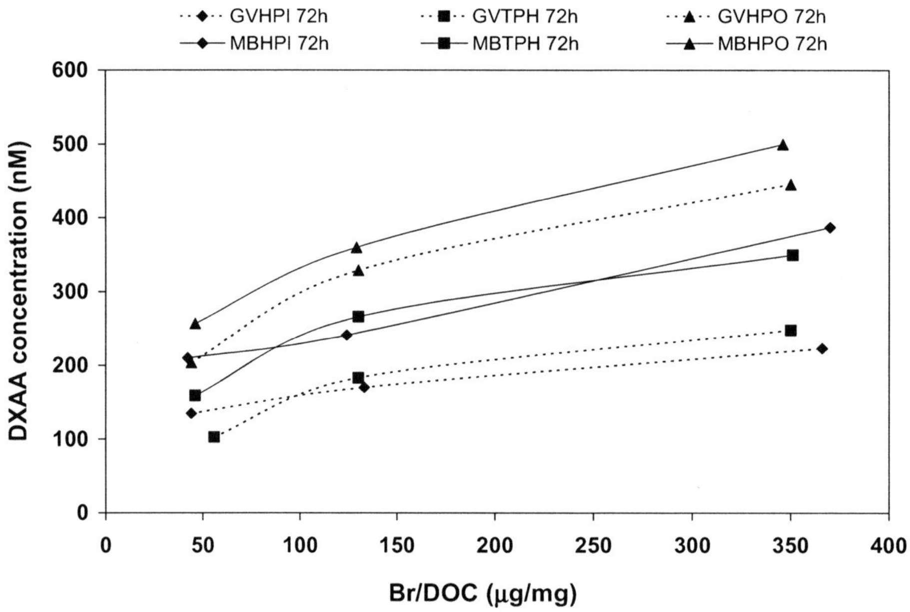

Figure 5.19 Effect of bromide on 72-hr DXAA formation for all fractions of Greenville and Myrtle Beach DOM at pH 6

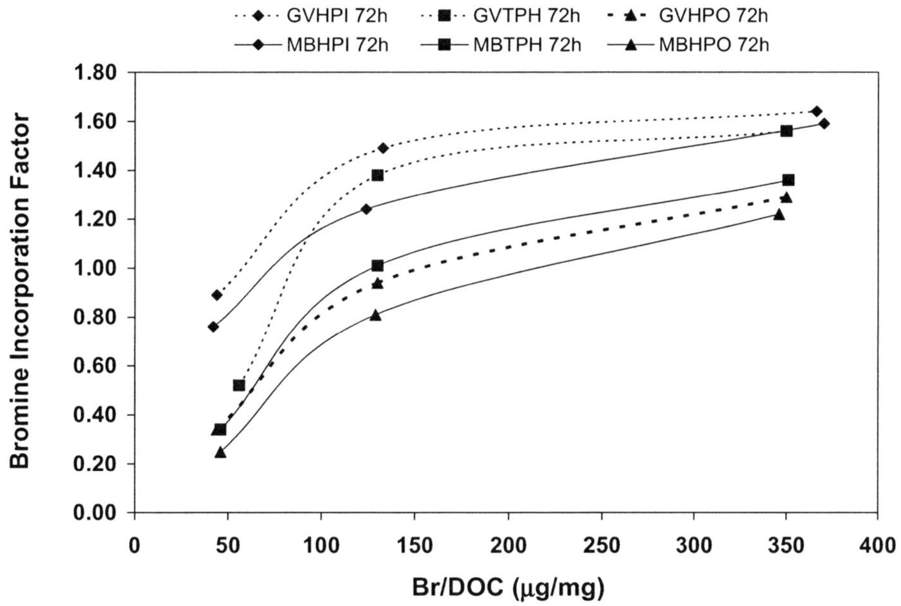

Figure 5.20 Effect of bromide on 72-hr bromine incorporation for all fractions of Greenville and Myrtle Beach DOM at pH 6

CHAPTER 6
CONCLUSIONS AND RECOMMENDATIONS

The important conclusions and recommendations from different phases of this study are compiled and summarized below.

RO ISOLATION AND RESIN FRACTIONATION OF DOM AND THE IMPACT ON HAA AND THM FORMATION

- RO isolation is an effective method for concentrating and isolating DOM from natural waters. Our experiences in this study and in our previous work have shown that higher DOM recoveries are obtained in waters having high SUVA and high DOC values than those having low SUVA and low DOC. This is attributed to generally increasing molecular weight of DOM components with increasing SUVA values. In this study, it was possible to recover 87, 88 and 99 of organic carbon from the source waters of Spartanburg (SP), Greenville (GV) and Myrtle Beach (MB) water treatment plants, respectively, during RO isolation.
- A hydrogen or sodium exchanger prior to an RO system can be used to remove multivalent cations from water. The hydrogen exchanger reduces the cation content (except hydrogen) in the RO concentrate. However, the use of a hydrogen exchanger significantly increases DOM losses (to both permeate and membrane) during RO concentration. This is due to the decreasing size of DOM and decreasing repulsive electrostatic interactions between the membrane surface and the DOM molecules with decreasing pH when a hydrogen exchanger is used. In this study, for GV water, DOM recoveries of 74 and 88% were obtained using hydrogen and sodium exchangers, respectively. The disadvantage of using a sodium exchanger is the presence of high sodium concentration in the RO concentrate, which can be problematic depending on the objective of subsequent use of the RO concentrate. Since the presence of high background sodium concentration was not an important factor in this study, a sodium exchanger was used in RO isolation.
- The DOM losses during RO isolation occur either through the permeate or to the membrane (i.e., fouling of the membrane). In this study, higher degrees of DOM losses through permeate were observed for low $SUVA_{254}$ than high $SUVA_{254}$ waters. This was attributed to the high molecular weight of DOM components in the latter than the former, as indicated by the $SUVA_{254}$ values. However, no correlation has been observed between the $SUVA_{254}$ values of the waters and the degree of losses to the membrane (characterized as NaOH wash). On the other hand, the total DOM losses (i.e., permeate and membrane) showed an increasing trend with decreasing $SUVA_{254}$ values of the raw water.
- "Reactivity check" experiments showed that there was no impact of RO isolation on the DOM reactivity toward THM and HAA formation and speciation. In addition, DBP formation of RO isolates did not change during the 2–4 months of storage period in a 4°C refrigerator in the dark.
- Since both organics and inorganics are concentrated during RO isolation, the concentration of inorganics, especially silica may be problematic during the

subsequent resin fractionation of DOM. Therefore, the concentration factor is an important consideration for RO isolation. In this study a high concentration factor (313) was used for the low DOC GV water to collect enough DOM mass from the source water. However, some degree of silica precipitation was observed as a result of adjusting the pH of RO concentrate to 2 prior to XAD-8 column fractionation. As a result, some of the TPH and HPI components in the GV water might have been captured in the HPO fraction since the adsorption of DOM by silica increases with decreasing pH. Silica was not a problem for the MB sample due to low concentration factor needed for its concentration.

- Resin fractionation results showed that GV and MB waters are composed of 61% HPO, 17% TPH, 22% HPI, and 72% HPO, 17% TPH, 11% HPI, respectively. The HPO fractions had the highest $SUVA_{254}$ values for both samples ($SUVA_{254,GV,HPO}$ = 3.3 and $SUVA_{254,MB,HPO}$ = 5.0 L/mg-m). While the TPH fraction had a $SUVA_{254}$ lower than the HPI fractions in GV water ($SUVA_{254,GV,TPH}$ = 1.2 and $SUVA_{254,GV,HPI}$ = 1.9 L/mg-m), the $SUVA_{254}$ values of MB TPH and HPI fractions were the same ($SUVA_{254,MB,TPH}$ = 2.2 and $SUVA_{254,MB,HPI}$ = 2.2 L/mg-m). Visual observations (e.g., a brown color in NaOH wash solutions) during resin cleaning after fractionation indicated that some DOM components were not completely recovered from the resins. Calculation of the $SUVA_{254}$ of the whole water from the individual $SUVA_{254}$ values of the fractions and comparison with the measured value of the whole water indicated that especially for MB water, high UV absorbing components were not completely recovered from the resins during fractionation.
- To quantify the impact of DOM losses during fractionation, reactivity check experiments were conducted side by side at the same DOC concentration for the original RO concentrate and the reconstituted RO concentrate prepared using the individual DOM fractions. The results using chlorine showed that the GV reconstituted water from the individual fractions yielded 12 and 23% lower THM and HAA formations, respectively, compared to those of RO concentrate. For the MB water, the THM formation of the RO concentrate and reconstituted water was similar under both chlorination and chloramination conditions, while a lower degree of HAA formation (14 and 20%) was observed in the reconstituted water than in the RO concentrate. These results indicated that the DOM components lost during fractionation were mostly HAA precursors. Despite some of these losses, the results also showed that a great majority (i.e., >77%) of DOM reactivity toward THM and HAA formation in the source waters was captured in the collected fractions.

HAA FORMATION PATHWAYS AND KINETICS DURING CHLORAMINATION AND PRACTICAL IMPLICATIONS

- It was found that the direct reaction of NH_2Cl with DOM is the major pathway for HAA formation during chloramination. Experiments performed in the presence (where NH_2Cl decomposition and $NHCl_2$ formation are significantly suppressed) and absence (where NH_2Cl decomposition can occur) of background ammonia indicated that a major amount (~80%) of DXAA was formed from the direct reaction(s) of NH_2Cl with DOM at pH 6. The remaining ~20% resulted from reactions with HOCl. Results showed that $NHCl_2$ did not form HAA under typical drinking water conditions. The contribution of HOCl to the overall DXAA formation decreased with increasing pH. The amount of DXAA formed from the direct reaction(s) significantly decreased with increasing pH. This result indicates that even if NH_2Cl decomposition is minimized or if NH_2Cl is formed by mixing Cl_2 and NH_3 at the same injection location, an appreciable amount of DXAA may still form with decreasing pH values.

- It was shown that HAA formation kinetics during chloramination are not extremely fast, as has been previously reported in the literature. It was discovered and reported for the first time that the main reason for the different patterns reported for HAA formation kinetics in the literature is the use of NH_4Cl as the quenching agent recommended by USEPA Method 552.3. Since NH_4Cl does not quench residual chloramine, the reactions of chloramine can continue until the samples are analyzed. As a result, false high HAA concentrations are measured at early reaction periods. The practical consequence of this finding is the possibility of overestimating HAA formation for distribution system samples close to the treatment plant. For one water tested in this study, it was found that for the samples within the first 8 hr of chloramination at pH 7 and 8.3, a positive error of up to 10 μg/L was observed in the measured HAA concentrations by using NH_4Cl as the quenching agent and storing the samples in a refrigerator for seven days. The magnitude of the error decreased with increasing contact time.
- In a limited number of experiments, five quenching approaches, stoichiometric amount of Na_2SO_3, NH_4Cl, sodium meta-arsenite ($NaAsO_2$), increasing pH to 10 using a borate acid buffer solution and increasing pH to 10 followed by 100 mg/L ascorbic acid (i.e., Vitamin C) addition, were tested side by side using the same chloraminated sample. $NaAsO_2$ produced the best results, showing no effect on all nine HAA determination even after 14 days of storage. Furthermore, the biocide capability of $NaAsO_2$ has been previously shown for distribution system samples. Unfortunately, due its toxic nature, the use of $NaAsO_2$ in practical application poses various hazard and waste handling issues. More research is needed to identify and validate a proper quenching approach for chloraminated DBP samples in practical applications. If a short storage time is needed (1–2 days) and if the residual chloramine concentration can be quickly determined in a sample, the use of a stoichiometric amount of Na_2SO_3 may be considered. However, it should be noted that Na_2SO_3 does not work as a biocide, thus degradation of HAA species may occur in biologically active samples. The use of excess Na_2SO_3 concentrations results in the decomposition of some brominated HAA species.
- The results also showed that chloramination is an effective way to control DBP formation, especially by minimizing the lag time between chlorine and ammonia addition and maintaining pH above 7.5. Since HAA formation was found in chloraminated samples to occur mainly from the direct reaction of NH_2Cl with DOM, the enhanced removal of DOM can decrease HAA formation during chloramination.

HAA FORMATION FROM DOM AND DOM FRACTIONS DURING CHLORAMINATION

- The results obtained in this study and in previous investigations with preformed chloramines indicate that at least 80% of HAA to be formed from chloramines(s) in natural waters will be in the DXAA form, and the contribution of DXAA may reach and exceed over 90% in some cases. Small amount of MXAA formation may also be observed, while formation of TXAA is usually negligible.

- *pH effect:* DXAA formation both on a molar and mass concentration basis decreased with increasing pH at all three bromide concentrations in both waters. Formation of MXAA, although in very small amounts, appeared to decrease with increasing pH, while TXAA formation was usually negligible, except small amounts of TXAA formation was observed at pH 6 with the highest Br/DOC ratios. These trends were different from those observed during chlorination. For chlorination, formations of MXAA and DXAA is almost independent of pH at typical water treatment conditions (i.e., pH 6 to 9), while formation of TXAA decreases with increasing pH. The formation of DXAA was very sensitive to pH during chloramination. The pH effect was not proportional to the change in pH. There was a larger decrease in DXAA yield when pH was increased from 6 to 7.5 than from 7.5 to 9.0. It is clear from these results that water utilities intending to decrease HAA formation during chloramination should maintain pH above 7.5.
- *Bromide effect:* Bromide effect was pH dependent. At pH 6, DXAA formation increased and shifted toward brominated species with increasing bromide concentration. Small amounts of MXAA formation were observed, which shifted also from MCAA to MBAA with increasing Br/DOC ratio. The presence of bromide resulted in TXAA formation at the highest Br/DOC ratios for both DOMs. At pH 7.5, the impact of bromide on DXAA formation and NH_2Cl decomposition was relatively small, and at pH 9, it was essentially negligible. Bromine incorporation was higher and occurred more rapidly at pH 6 than pH 7.5. No brominated specie was formed at pH 9. These results showed that one additional advantage of having an operational pH of 7.5 or above for chloramination is to reduce the formation of brominated HAA species.
- *DOM effect:* The aromatic content of DOM (i.e., $SUVA_{254}$) appeared to have some impact on the DXAA yields during chloramination at pH values (e.g., pH 6 tested in this study) that considerably increase NH_2Cl decomposition. However, the effect of $SUVA_{254}$ on the HAA yield is much less pronounced during chloramination than chlorination. In this study, at pH 6 the formation of DXAA in MB DOM ($SUVA_{254}$ = 4.9 mg/L-m) was about 10–30% higher (on both mass and molar basis) than in GV ($SUVA_{254}$ = 2.6 mg/L-m) and SP DOM ($SUVA_{254}$ = 3.2 mg/L-m) at three bromide levels. At pH 7.5 and 9.0, both GV and MB DOMs showed very similar DXAA yields. These results indicated that pH was a more important factor controlling the extent of DXAA formation and speciation than $SUVA_{254}$ of DOM and bromide in typical drinking water conditions.
- *Reactivity of DOM fractions:* To investigate the effect of DOM characteristics, two surface waters with different $SUVA_{254}$ values were fractionated into hydrophobic (HPO), transphilic (TPH), and hydrophilic (HPI) fractions. Generally, with the decrease in pH, the effect of DOM characteristics and bromide became more important. At pH 9, the effect of DOM was minimal and the effect of bromide was not observed. At pH 7.5, both effects were observed but not to a significant degree. At pH 6, DXAA yield increased with increasing $SUVA_{254}$ for all fractions. However, the effect of $SUVA_{254}$ on the HAA yields was less pronounced during chloramination than during chlorination. The difference in DXAA yields between chlorination and

chloramination increased with increase in $SUVA_{254}$. The overall trend of DXAA formation was (MBHPO > GVHPO) > (MBTPH ≈ MBHPI) > (GVTPH ≈ GVHPI), indicating that the hydrophobic components were the major precursors of DXAA. The opposite trend held true for the bromine incorporation factor (n): GVHPI ≈ GVTPH > MBHPI > MBTPH ≈ GVHPO > MBHPO. The bromine incorporation factor showed an inverse but not a linear correlation with $SUVA_{254}$.

- The results obtained in this phase of the study indicates that utilities need to maintain the pH above 7.5 during operation to control DBP formation during chloramination, which is also helpful to maintain a relatively stable NH_2Cl residual in a distribution system. Although the HPI fractions were more reactive with bromide than the HPO fractions, the HPO fractions consistently showed higher DXAA yields than the relatively hydrophilic fractions. Thus, the enhanced removal of the hydrophobic component especially in moderate and high $SUVA_{254}$ waters can assist with the better control of DBP formation during chloramination. Therefore, chloramination is expected to benefit from (enhanced) coagulation/flocculation which has been shown to preferentially remove hydrophobic DOM components.

FUTURE RESEARCH NEEDS

The experiments in this study were performed using the water samples collected from the influents of three water treatment plants. Future studies should be conducted using the water samples after treatment processes prior to any disinfectant addition and water distribution system samples to further confirm and expand the findings regarding the HAA formation routes and quenching agent effect, respectively. More research is needed to identify and validate a proper quenching approach for chloraminated DBP samples in practical applications. There is wealth of HAA formation kinetics data obtained in this study that can be modeled with the DBP formation models that were developed for chloramination in the recent years. Furthermore, new and improved mechanistic HAA formation models can be developed using the information about the HAA formation routes obtained in this study.

REFERENCES

Aiken, G.R., and E. Cotsaris. 1995. Soil and Hydrology: Their Effect on NOM. *Journal American Water Works Association*, 87(1):36–45.

Aiken, G.R., D.M. McKnight, K.A. Thorn, and E.M. Thurman. 1992. Isolation of Hydrophilic Organic Acids From Water Using Nonionic Macroporous Resins. *Organic Geochemistry*, 18(4):567–573.

Aiken, G.R., E.M. Thurman, and R.L. Malcolm. 1979. Comparison of XAD Macroporous Resins for the Concentration of Fulvic Acid From Aqueous Solution. *Analytical Chemistry*, 51(11):1799–1803.

American Water Works Association Committee Report. 2000. Disinfection at Large and Medium-Size Systems. *Journal American Water Works Association*, 92(5):32–43.

Amy, G., M. Siddiqui, W. Zhai, J. Bebroux, and W. Odem. 1994. *Survey of Bromide in Drinking Water and Impacts on DBP Formation*. Denver, Colo.: AwwaRF and AWWA.

Amy, G.L., L. Tan, and M.K. Davis. 1991. The Effects of Ozonation and Activated Carbon Adsorption on Trihalomethane Speciation. *Water Research*, 25(2):191–202.

APHA, AWWA, and WEF (American Public Health Association, American Water Works Association, and Water Environment Federation). 1998. *Standard Methods for the Examination of Water and Wastewater*, 20th ed. Washington, D.C.: APHA.

Baribeau, H., S.W. Krasner, R. Chinn, and P.C. Singer. 2005. Impact of Biomass on the Stability of HAAs and THMs in a Simulated Distribution System. *Journal American Water Works Association*, 97(2):69–81.

Bousher, A., P. Brimblecombe, and D. Midgley. 1986. Rate of Hypobromite Formation in Chlorinated Seawater. *Water Research*, 20(7):865–870.

Bousher, A., P. Brimblecombe, and D. Midgley. 1989. Kinetics of Reactions in Solutions Containing Monochloramine and Bromide. *Water Research*, 23(8):1049–1058.

Brodtmann, N.V., and P.J. Russo. 1979. The Use of Chloramine for Reduction of Trihalomethanes and Disinfection of Drinking Water. *Water Technology*, 79(1):40–42.

Brophy, K.S., H.S. Weinberg, and P.C. Singer. 2000. Quantification of Nine Haloacetic Acids Using Gas Chromatography With Electron Capture Detection. In *Natural Organic Matter and Disinfection By-Products: Characterization and Control in Drinking Water*. Edited by S.E. Barrett, S.W. Krasner, and G.L. Amy. ACS Symposium Series No 761. Washington, D.C.: American Chemical Society.

Brown, M. 1975. High Molecular Weight Material in Baltic Seawater. *Marine Chemistry*, 3:253–258.

Bull, R.J., S.W. Krasner, P.A. Daniel, and R.D. Bull. 2001. *Health Effects and Occurrence of Disinfection By-Products*. Denver, Colo.: AwwaRF and AWWA.

Burttschell, R.H., A.A. Rosen, F.M. Middleton, and M.B. Ettinger. 1959. Chlorine Derivatives of Phenol and Phenol Causing Taste and Odor. *Journal American Water Works Association*, 51(2):205–214.

Chang, E.E., Y.P. Ling, and P.C. Chiang. 2001. Effects of Bromide on the Formation of THMs and HAAs. *Chemosphere*, 43(8):1029–1034.

Chin, Y.P., G. Aiken, and E. OÕLoughlin. 1994. Molecular Weight, Polydispersity, and Spectroscopic Properties of Aquatic Humic Substances. *Environmental Science and Technology*, 28(11):1853–1858.

Chow, A.T., F. Guo, S. Gao, and R.S. Breuer. 2006. Size and XAD Fractionations of Trihalomethane Precursors From Soils. *Chemosphere*, 62:1636–1646.

Christman, R.F., D.L. Norwood, D.S. Millington, J.D. Johnson, and A.A Stevens. 1983. Identity and Yields of Major Halogenated Products of Aquatic Fulvic Acid Chlorination. *Environmental Science and Technology*, 17(10):625–628.

Collins, M.R., G.L Amy, and C. Steelink. 1986. Molecular Weight Distribution, Carboxylic Acidity and Humic Substances Content of Aquatic Organic Matter: Implications for Removal During Drinking Water Treatment. *Environmental Science and Technology*, 20(10):1028–1032.

Cowman, G.A, and P.C. Singer. 1996. Effect of Bromide Ion on Haloacetic Acid Speciation Resulting From Chlorination and Chloramination of Aquatic Humic Substances. *Environmental Science and Technology*, 30(1):16–24.

Croué, J.P., M.F. Benedetti, D. Violleau, and J.A. Leenheer. 2003. Characterization and Copper Binding of Humic and Nonhumic Organic Matter Isolated From the South Platte River: Evidence for the Presence of Nitrogenous Binding Site. *Environmental Science and Technology*, 37(2):328–336.

Croué, J.P., D. Violleau, and L. Labouyrie. 2000. Disinfection By-Product Formation Potentials of Hydrophobic and Hydrophilic Natural Organic Matter Fractions: A Comparison Between a Low- and High-Humic Water. In *Natural Organic Matter and Disinfection By-Products: Characterization and Control in Drinking Water.* Edited by S.E. Barrett, S.W. Krasner, and G.L. Amy. ACS Symposium Series No 761. Washington, D.C.: American Chemical Society.

Diehl, A.C., G.E. Speitel, J.M. Symons, S.W. Krasner, C.J. Hwang, and S.E. Barrett. 2000. DBP Formation During Chloramination. *Journal American Water Works Association*, 92(6):76–90.

Duirk, S.E. 2003. Mechanisms and Modelling of Monochloramine Loss, Natural Organic Matter Oxidation, and Disinfection By-Products Formation in Water. Ph.D. diss., University of Iowa, Iowa City.

Edwards, M., and A. Dudi. 2004. Role of Chlorine and Chloramine in Corrosion of Lead-Bearing Plumbing Materials. *Journal American Water Works Association*, 96(10):69–81.

Fairhurst, A.J., P. Warwick, and S. Richardson. 1995. The Influence of Humic Acid on the Adsorption of Europium Onto Inorganic Colloids as a Function of pH. *Colloids and Surfaces A: Physicochemical and Engineering Aspects*, 99:187–199.

Gazda, M., L.E. Dejarem, T.K. Choudhury, R.G. Cooks, and D.W. Margerum. 1993. Mass Spectrometric Evidence for the Formation of Bromochloramine and N-Bromo-N-Chloromethylamine in Aqueous Solution. *Environmental Science and Technology*, 27(3):557–561.

Gazda, M., and D.W. Margerum. 1994. Reactions of Monochloramine with Br_2, Br_3^-, HOBr, and OBr^-: Formation of Bromochloramines. *Inorganic Chemistry*, 33(1):118–123.

Ghosh, K., and M. Schnitzer. 1980. Macromolecular Structure of Humic Substances. *Soil Science*, 129:266–276.

Glaser, J.A., D.L. Foerst, G.D. McKee, S.A. Quave, and W.L. Budde. 1981. Trace Analyses for Wastewater. *Environmental Science and Technology*, 15(12):1426–1435.

Glaze, W.H. 1990. Group Report: What Is the Composition of Organic Acids in Aquatic Systems and How Are They Characterized? In *Organic Acids in Aquatic Ecosystems.* Edited by E.M. Perdue and E.T. Gjessing. New York: John Wiley & Sons.

Gray, E.T., D.W. Margerum, and R.P. Huffman. 1978. Chloramine Equilibria and the Kinetics of Disproportionation in Aqueous Solution. In *Organometals and Organometalloids: Occurrence and Fate in the Environment.* Edited by F.E. Brinckman and J.M. Bellama. Washington, D.C.: American Chemical Society.

Greyshock, A.E., and P.J. Vikesland. 2006. Triclosan Reactivity Chloraminated Waters. *Environmental Science and Technology*, 40(8):2615–2622.

Haas, C.N. 1999. Disinfection. In *Water Quality and Treatment: A Handbook of Community Water Supply.* Edited by R.D. Letterman. New York: McGraw-Hill.

Hand, V.C., and D.W. Margerum. 1983. Kinetics and Mechanisms of the Decomposition of Dichloramine in Aqueous Solution. *Anaytical Chemistry*, 22(10):1449–1456.

Hayes, M.H.B., P. MacCarthy, R.L. Malcolm, and R.S. Swift, eds. 1989. Structures of Humic Substances: The Emergence of Forms. In *Humic Substances II: In Search of Structure.* New York: John Wiley & Sons.

Heasley, V.L., A.M. Fisher, E.E. Herman, F.E. Jacobsen, E.W. Miller, A.M. Ramirez, N.R. Royer, J.M. Whisenand, D.L Zoetewey, and D.F. Shellhamer. 2004. Investigations of the Reactions of Monochloramine and Dichloramine With Selected Phenols: Examination of Humic Acid Models and Water Contaminants. *Environmental Science and Technology*, 38(19):5022–5029.

Hong, Y., S. Liu, H. Song, and T. Karanfil. 2005a. HAA Formation During Chloramination. In *Proc. of the 2005 Annual Conference.* Denver, Colo.: AWWA.

Hong, Y., S. Liu, H. Song, and T. Karanfil. 2005b. The Role of NH_2Cl and HOCl in the Formation of Haloacetic Acids During Chloramination. Presented at 230th National Meeting of American Chemical Society in Washington, D.C., Division of Environmental Chemistry, Extended Abstracts and Program, Vol.45, No. 2.

Hua, G., and D.A. Reckhow. 2005a. Factors Affecting the Formation of Disinfection By-Product During Chlorination and Chloramination. In *Proc. of the AWWA Water Quality Technology Conference.* Denver, Colo.: AWWA.

Hua, G., and D.A. Reckhow. 2005b. Characterization of Disinfection By-Product Precursors Based on Hydrophobicity and Molecular Size. In *Proc. of the AWWA Annual Conference.* Denver, Colo.: AWWA.

Hwang, C.J., M.J. Sclimenti, A. Bruchet, J.P. Croué, and G.L. Amy. 2001. DBP Yields of Polar NOM Fractions From Low-Humic Waters. In *Proc. of the Water Quality Technology Conference.* Denver, Colo.: AWWA.

Hwang, C.J., M.J. Sclimenti, and S.W. Krasner. 2000. Disinfection By-Product Formation Reactivities of Natural Organic Matter Fractions of a Low-Humic Water. In *Natural Organic Matter and Disinfection By-Products: Characterization and Control in Drinking Water.* Edited by S.E. Barrett, S.W. Krasner, and G.L. Amy. ACS Symposium Series No 761. Washington, D.C.: American Chemical Society.

Jafvert, C.T., and R.L. Valentine. 1987. Dichloramine Decomposition in the Presence of Excess Ammonia. *Water Research*, 21(8):967–973.

Jafvert, C.T., and R.L. Valentine. 1992. Dichloramine Decomposition in the Presence of Excess Ammonia. *Environmental Science and Technology*, 26(3):577–586.

Jensen, J.N., J.D. Johnson, J.J. St. Aubin, and R.F. Christman. 1985. Effect of Monochloramine on Isolated Aquatic Fulvic Acid. *Organic Geochemistry*, 8(1):71–76.

Johnson, J.D., and J.N. Jensen. 1986. THM and TOX Formation: Routes, Rates and Precursors. *Journal American Water Works Association*, 78(4):156–162.

Kilduff, J.E., S. Mattaraj, A. Wigton, M. Kitis, and T. Karanfil. 2004. Effects of Reverse Osmosis Isolation on Reactivity of Naturally Occurring Dissolved Organic Matter in Physicochemical Processes. *Water Research*, 38(4):1026–1036.

Kilduff, J.E., and W.J. Weber Jr. 1992. Transport and Separation of Organic Macromolecules in Ultrafiltration Processes. *Environmental Science and Technology*, 26,:69–577.

Kitis, M., T. Karanfil, A. Wigton, and J.E. Kilduff. 2002. Probing the Reactivity of Dissolved Organic Matter for Disinfection By-Product Formation Using XAD8 Resin Adsorption and Ultrafiltration Fractionation. *Water Research*, 36(10):3834–3848.

Kitis, M., J.E. Kilduff, and T. Karanfil. 2001. Isolation of Dissolved Organic Matter (DOM) From Surface Waters Using Reverse Osmosis and Its Impact on the Subsequent Reactivity of DOM to Formation and Speciation of Disinfection By-Products. *Water Research*, 35(9):2225–2234.

Korshin, G.V., M.M. Benjamin, and R.S. Sletten. 1997a. Adsorption of Natural Organic Matter (NOM) on Iron Oxide: Effects on NOM Composition and Formation of Organo-halide Compounds During Chlorination. *Water Research*, 31(7):1643–1650.

Korshin, G.V., C.W. Li, and M.M. Benjamin. 1997b. Monitoring the Properties of Natural Organic Matter Through UV Spectroscopy: A Consistent Theory. *Water Research*, 31(7):1787–1795.

Krasner, S.W., and S.E. Barrett. 1985. Aroma and Flavor Characteristics of Free Chlorine and Chloramines. In *Proc. of the 1984 AWWA Water Quality Technology Conference*. Denver, Colo.: AWWA.

Krasner, S.W., J.P. Croué, J. Buffle, and E.M. Perdue. 1996a. Three Approaches for Characterizing NOM. *Journal American Water Works Association*, 88(6):66–79.

Krasner, S.W., M.J. McGuire, J.G. Jacangelo, N.L. Patania, K.M. Reagan, and E.M. Aieta. 1989. The Occurrence of Disinfection By-Products in U.S. Drinking Water. *Journal American Water Works Association*, 81(8):41–53.

Krasner, S.W., M.J. Sclimenti, R. Chinn, Z.K. Chowdhury, and D.M. Owen. 1996b. The Impact of TOC and Bromide on Chlorination By-Product Formation. In *Disinfection By-Products in Water Treatment: The Chemistry of Their Formation and Control*. Edited by R.A. Minear and G. Amy. Boca Raton, Fla.: CRC Lewis Publishers.

Krasner, S.W., M.J. Sclimenti, and E.G. Means. 1994. Quality Degradation: Implication for DBP Formation. *Journal American Water Works Association*, 86(6):34–47.

Kwak, J.C.T., and R.W.P. Nelson. 1977. Ultrafiltration of Fulvic and Humic Acids: A Comparison of Stirred Cell and Hallow Fiber Techniques. *Geochemica et Cosmochimica Acta*, 41:993–996.

Leao, S.F. 1981. Kinetics of Combined Chlorine: Reactions of Substitution and Redox. Ph.D. diss., University of California, Berkeley.

Leenheer, J.A. 1981. Comprehensive Approach to Preparative Isolation and Fractionation of Dissolved Organic Carbon From Natural Waters and Wastewaters. *Environmental Science and Technology*, 15(5):578–587.

Leenheer, J.A., and J.P. Croué. 2003. Characterizing Aquatic Dissolved Organic Matter. *Environmental Science and Technology*, 37(1):18A–27A.

Leenheer, J.A., J.P. Croué., M. Benjamin, G.V. Korshin, C.J. Hwang, A. Bruchet, and G.R. Aiken. 2000. Comprehensive Isolation of Natural Organic Matter From Water for Spectral Characterization and Reactivity Testing. In *Natural Organic Matter and Disinfection By-Products: Characterization and Control in Drinking Water.* Edited by S.E. Barrett, S.W. Krasner, and G.L. Amy. ACS Symposium Series No 761. Washington, D.C.: American Chemical Society.

Leung, S.W., and R.L. Valentine. 1994. An Unidentified Chloramine Decomposition Product I. Chemistry and Characterization. *Water Research*, 28(6):1475–1483.

Li, C.W., M.M. Benjamin, and G.V. Korshin. 2000. Use of UV Spectroscopy to Characterize the Reaction Between NOM and Free Chlorine. *Environmental Science and Technology*, 34(12):2570–2575.

Liang, L., and P.C. Singer. 2003. Factors Influencing the Formation and Relative Distribution of Haloacetic Acids and Trihalomethanes in Drinking Water. *Environmental Science and Technology*, 37(13):2920–2928.

Long, B.W., G.S. Longsdon, and D.G. Neden. 1992. Options for Controlling Disinfection Byproducts in Greater VancouverÕs Water Supply. In *Proc. of the Water Quality Technology Conference, Part I.* Denver, Colo.: AWWA.

MacCarthy, P., R.L. Malcolm, C.E. Clapp, and P.R. Bloom. 1990. An Introduction to Soil Humic Substance. In *Humic Substances in Soil and Crop Sciences: Selected Readings.* Edited by P. MacCarthy, C.E. Clapp, R.L. Malcolm, and P.R. Bloom. Madison, Wis.: American Society of Agronomy and Soil Science Society of America.

Malcolm, R.L. 1991. Factors to Be Considered in the Isolation and Characterization of Aquatic Humic Substances. In *Humic Substances in the Aquatic and Terrestrial Environment.* Edited by B. Allard, H. Boren, and A. Grimvall. Berlin: Springer-Verlag.

Malcolm, R.L., and P. MacCarthy. 1992. Quantitative Evaluation of XAD8 and XAD4 Resins Used in Tandem for Removing Organic Solutes From Water. *Environment International*, 18(6):597–607.

Margerum, D.W., E.T. Gray Jr., and R.P. Huffman. 1978. Chlorination and the Formation of N-Chloro Compounds in Water Treatment. In *Organometals and Organometalloids: Occurrence and Fate in the Environment.* Edited by F.E. Brinkman and J.M. Bellama. Washington, D.C.: ACS Books.

McGuire, M.J., and R.G. Meadow. 1988. AwwaRF Trihalomethane Survey. *Journal American Water Works Association*, 80(1):61–68.

Mitch, W.A., J.O. Sharp, R.R. Trussell, R.L. Valentine, L. Alvarez-Cohen, and D.L. Sedlak. 2003. *N*-Nitrosodimethylamine as a Drinking Water Contaminant: A Review. *Environmental Engineering Science*, 20(5):389–404

Morris, J.C., and R.A. Isaac. 1981. A Critical Review of Kinetic and Thermodynamic Constants for the Aqueous Chlorine-Ammonia System. In *Water Chlorination: Environmental Impact and Health Effects.* Edited by R.L. Jolley, W.A. Brungs, J.A. Cotruvo, R.B. Cumming, J.S. Mattice, and V.A. Jacobs. Ann Arbor, Mich.: Ann Arbor Science.

Najm, I.N., N.L. Patania, J.G. Jacangelo, and S.W. Krasner. 1994. Evaluating Surrogates for Disinfection By-Products. *Journal American Water Works Association*, 94(6):98–106.

Noot, D.F., W.B. Anderson, S.A. Daignault, D.T. Williams, and P.M. Huck. 1989. Evaluating Treatment Process With Ames Mutagennicity. *Journal American Water Works Association*, 81(9):87–102.

Norman, T., L. Harms, and R. Looyenga. 1980. The Use of Chloramines to Prevent Trihalomethane Formation. *Journal American Water Works Association*, 72(3):176–180.

Pepich, B.V., M.M. Domino, R.A. Datillo, P.S. Fair, and D.J. Munch. 2004. Validating Sample Preservation Techniques and Holding Times for the Approved Compliance Monitoring Methods for Haloacetic Acids Under the USEPAÕs Stage I D/DBP Rule. *Water Research*, 38(4):895–902.

Pope, P.G., G.E. Speitel, M. Martin-Doole, and M.R. Collins. 2002. Factors Contributing to DXAA Formation During Chloramination. In *Proc. of the Water Quality and Technology Conference*. Denver, Colo.: AWWA.

Pope, P.G., G.E. Speitel, M. Martin-Doole, and M.R. Collins. 2003. DXAA Formation Kinetics During Chloramination. In *Proc. of the AWWA Annual Conference*. Denver, Colo.: AWWA.

Qi, Y., C. Shang, and M.C. Lo. 2004. Formation of Haloacetic Acids During Monochloramination. *Water Research*, 38(9):2375–2383.

Reckhow, D.A., and P.C. Singer. 1985. Mechanisms of Oragnic Halide Formation During Fulvic Acid Chlorination and Implications With Respect to Preozonation. In *Water Chlorination: Chemistry, Environmental Impact and Health Effects*, Vol. 5. Chelsea, Mich.: Lewis Publishers.

Reckhow, D.A., P.C. Singer, and R.L. Malcolm. 1990. Chlorination of Humic Materials: By-Product Formation and Chemical Interpretations. *Environmental Science and Technology*, 24(11):1655–1664.

Seidel, C., M.J. McGuire, R.S. Summers, and S. Via. 2005. Have Utilities Switched to Chloramines? *Journal American Water Works Association*, 97(10):87–97.

Siddiqui, M., and G.L. Amy. 1993. Factors Affecting DBP Formation During Ozone-Bromide Reactions. *Journal American Water Works Association*, 85(1):63–72.

Singer, P.C., and S.D. Chang. 1989. Correlations Between Trihalomethanes and Total Organic Halides Formed During Drinking Water Treatment. *Journal American Water Works Association*, 81(8):61–65.

Singer, P.C., and G. Harrington. 1993. Coagulation of DBP Precursors: Theoretical and Practical Considerations. In *Proc. of the Water Quality Technology Conference*. Denver, Colo.: AWWA.

Singer, P.C., G. Harrington, G. Cowman, M. Smith, D. Schechter, and L. Harrington. 1999. *Impacts of Ozonation on the Formation of Chlorination and Chloramination By-Products*. Denver, Colo.: AwwaRF and AWWA.

Singer, P.C., H.S. Weinberg, K. Brophy, L. Liang, M. Roberts, I. Grisstede, S. Krasner, H. Baribeau, H. Arora, and I. Najm. 2002. *Relative Dominance of Haloacetic Acids and Trihalomethanes in Treated Drinking Water*. Denver, Colo.: AwwaRF and AWWA.

Sinha, S., G.L. Amy, and J. Sohn. 1997. Reactivity of NOM Fractions in Forming Chlorinated DBPs. In *Proc. of the American Water Works Association Annual Conference*. Denver Colo.: AWWA.

Snoeyink, V.L., and D. Jenkins. 1980. *Water Chemistry*. New York: John Wiley & Sons.

Speitel, G.E., P.G. Pope, M.R. Collins, and M. Martin-Doole. 2004. *Disinfection By-Product Formation and Control During Chloramination*. Denver, Colo.: AwwaRF and AWWA.

Stevens, A.A., L.A. Moore, and R.J. Miltner. 1989. Formation and Control of Non-Trihalomethane Disinfection By-Products. *Journal American Water Works Association*, 81(8):54–60.

Stumm, W., and J.J. Morgan. 1996. *Chemical Equilibria and Rates in Natural Waters*, 3rd ed. New York: John Wiley & Sons.

Summers, R.S., S.M. Hooper, H.M. Shukairy, G. Solarik, and D. Owen. 1996. Assessing DBP Yield: Uniform Formation Conditions. *Journal American Water Works Association*, 88(6):80–93.

Symons, J.M., G.E. Speitel, C.J. Hwang, S.W. Krasner, S.E. Barrett, S.C. Diehl, and R. Xia. 1998. *Factors Affecting Disinfection By-Product Formation During Chloramination*. Denver, Colo.: AwwaRF and AWWA.

Symons, J.M., R. Xia, G. Speitel, A. Diehl, C. Hwang, S. Krasner, and S. Barett. 1999. *Factors Affecting Disinfection By-Product Formation During Chloramination*. Denver, Colo.: AwwaRF and AWWA.

Tadanier, C.J., D.F. Berry, and W.R. Knocke. 1999. Dissolved Component Recovery Following Resin Exchange Based DOM Fractionation. *Journal of Environmental Engineering*, 125:933–943.

Teng, H., and J.N. Veenstra. 1996. A Study of Disinfection By-Products Formed Using Four Alternative Disinfectants as a Function of Precursor Characteristics. In *Disinfection By-Products in Water Treatment: The Chemistry of Their Formation and Control*. Edited by S.E. Barrett, S.W. Krasner, and G.L. Amy. ACS Symposium Series No 761. Washington, D.C.: American Chemical Society.

Thurman, E.M. 1985. *Organic Geochemistry of Natural Waters*. Dordrecht, Netherlands: Martinus Nijhoff/Dr. Junk Publishers.

Thurman, E.M., and R.L. Malcolm. 1981. Preparative Isolation of Aquatic Humic Substances. *Environmental Science and Technology*, 15(4):463–466.

Topudurti, K.V., and C.N. Haas. 1991. THM Formation by the Transfer of Active Chlorine From Monochloramine to Chloroacetophenone. *Journal American Water Works Association*, 83(5):62–66.

Town, R.M., and H.K.J. Powell. 1993. Limitations of XAD Resins for the Isolation of the Noncolloidal Humic Fraction in Soil Extracts and Aquatic Samples. *Analytica Chimica Acta* 271:195–202.

Trofe, T.W., G.W. Inman Jr., and J.D. Johnson. 1980. Kinetics of Monochloramine Decomposition in the Presence of Bromide. *Environmental Science and Technology*, 14(5):544–549.

USEPA (U.S. Environmental Protection Agency). 1996. Reprints of EPA Methods for Chemical Analyses Under the Information Collection Rule. EPA 814-B96-006. Cincinnati: USEPA.

USEPA (U.S. Environmental Protection Agency). 1998. National Primary Drinking Water Regulations: Disinfectants and Disinfection Byproducts; Final Rule. *Federal Register*, 63(241):69390. December 16, 1998.

USEPA (U.S. Environmental Protection Agency). 1999. Chloramines. In *Alternative Disinfectants and Oxidants Guidance Manual*. EPA 815-R999-014. Washington, D.C.: U.S. Environmental Protection Agency, Office of Water.

USEPA (U.S. Environmental Protection Agency). 2003. Determination of Haloacetic Acids and Dalapon in Drinking Water by Liquid-Liquid Microextraction, Derivatization, and Gas Chromatography With Electron Capture Detection. http://www.epa.gov.

USEPA (U.S. Environmental Protection Agency). 2006. National Primary Drinking Water Regulations: Stage 2 Disinfectants and Disinfection Byproducts Rule; Final Rule. 40 CFR Part 9, 141 and 142. *Federal Register* (Part II), 71(2):388. January 4, 2006.

Valentine, R.L. 1986. Bromochloramine Oxidation of N,N-Diethyl-p-Phenylenediamine in the Presence of Monochloramine. *Environmental Science and Technology*, 20(2):166–170.

Valentine, R.L., K.I. Brandt, and C.T. Jafvert. 1986. A Spectrophotometric Study of the Formation of an Unidentified Monochloramine Decomposition Product. *Water Research*, 20(8):1067–1074.

Valentine, R.L., and C.T. Jafvert. 1988. General Acid Catalysis of Monochloramine Disproportionation. *Environmental Science and Technology*, 22(6):691–696.

Valentine, R.L., K. Ozekin, and P.J. Vikesland. 1998. *Chloramine Decomposition in Distribution System and Model Waters*. Denver, Colo.: AwwaRF and AWWA.

Vikesland, P.J., K. Ozekin, and R.L. Valentine. 1998. Effect of Natural Organic Matter on Monochloramine Decomposition: Pathway Elucidation Through the Use of Mass and Redox Balances. *Environmental Science and Technology*, 32(10):1409–1416.

Vikesland, P.J., K. Ozekin, and R.L. Valentine. 2001. Monochloramine Decay in Model and Distribution System Waters. *Water Research*, 35(7):1766–1776.

Wajon, J.E., and J.C. Morris. 1980. Bromamination Chemistry: Rates of Formation of NH_2Br and Some N-Bromamino Acids. In *Water Chlorination: Chemistry, Environmental Impacts and Health Effects*, Vol. 3. Edited by R.L. Jolley, W.A. Brungs, R.B. Cumming, and V.A. Jacobs. Ann Arbor, Mich.: Ann Arbor Science.

White, G.C. 1999. *Handbook of Chlorination and Alternative Disinfectants,* 4th ed. New York: John Wiley & Sons.

Wu, W., P.A. Chadik, and J.D. Joseph. 2003. The Relationship Between Disinfection By-Product Formation and Structural Characteristics of Humic Substances in Chloramination. *Environmental Toxicology and Chemistry*, 22(12):2845–2852.

Zhang X., S. Echigo, R.A. Minear, and M.J Plewa. 2000. Characterization and Comparison of Four Major Disinfectants. In *Natural Organic Matter and Disinfection By-Products: Characterization and Control in Drinking Water.* Edited by S.E. Barrett, S.W. Krasner, and G.L. Amy. ACS Symposium Series No 761. Washington, D.C.: American Chemical Society.

ABBREVIATIONS

BAA	bromoacetic acid
BCAA	bromochloroacetic acid
BDCAA	bromodichloroacetic acid
BDCM	bromodichloromethane
Br	bromide
Br/DOC	bromide to dissolved organic carbon ratio
C	concentration
C_o	DOM stock concentration
CAA	chloroacetic acid
ClArOH	halogenated aromatic organic matter
Cl_2	chlorine
Cl_2/N	chlorine to nitrogen ratio
DBAA	dibromoacetic acid
DBCAA	dibromochloroacetic acid
DBCM	dibromochloromethane
DBP	disinfection by-product
DCAA	dichloroacetic acid
D/DBP	disinfectants/disinfectant by-products
D/DBPR	Disinfectants/Disinfection By-Products Rule
DDW	distilled deionized water
DL	detection limit
DN	dissolved nitrogen
DOC	dissolved organic carbon
DOM	dissolved organic matter
DOX	dissolved organic halides
DPD	N,N-diethyl-p-phenylenediamine
DXAA	sum of dihalogenated haloacetic acids
ECD	electron capture detector
G	mean velocity gradient
GC	gas chromatograph
GV	Greenville, South Carolina
HAA	haloacetic acid
HAA5	sum of five haloacetic acids (MBAA, MCAA, DBAA, DCAA, TCAA)
HAA6	sum of six HAA species containing Br (MBAA, DBAA, TBAA, BCAA, BDCAA, and DBCAA)
HAA_9	sum of nine haloacetic acids
HCl	hydrochloric acid
HOBr	hypobromous acid

HOCl	hypochlorous acid
HPI	hydrophilic fraction
HPO	hydrophobic fraction
ICP/MS	inductively coupled plasma/mass spectroscopy
IDSE	initial distribution system evaluation
k'	column capacity factor
MB	Myrtle Beach, South Carolina
MCL	maximum contaminant level
MRL	minimum reporting level
MTBE	methyl-tert butyl ether
MW	molecular weight
MXAA	monohalogenated haloacetic acids
N_2	nitrogen gas
NaOH	sodium hydroxide
Na_2SO_3	sodium sulfite
NBr_2Cl	dibromochloramine
NCl_3	trichloramine
NHBrCl	bromochloramine
$NHBr_2$	dibromomanie
NH_2Cl	monochloramine
$NHCl_2$	dichloramine
NH_3	ammonia
NH_4Cl	ammonium chloride
NOM	natural organic matter
OCl^-	hypochlorite ion
PAC	Project Advisory Committee
RAC	resin adsorption chromatography
RO	reverse osmosis
RSD	relative standard deviation
SiO_2	silica
SM	standard methods
SP	Spartanburg, South Carolina
SUVA	specific ultraviolet absorbance
$SUVA_{254}$	specific ultraviolet absorbance at 254 nm wavelength

t	contact time
TBAA	tribromoacetic acid
TBM	tribromomethane
TCAA	trichloroacetic acid
TCM	trichloromethane (chloroform
THM	total trihalomethanes
TOCl	total organic chlorine
TOX	total organic halides
TPH	transphilic fraction
TXAA	trihalogenated haloacetic acid
UFC	uniform formation potential conditions
USEPA	United States Environmental Protection Agency
UV	ultraviolet
V	volume
WTP	water treatment plant

Printed in the United States
137657LV00009B/4/P